From Quantum Information to Musical Semantics

From Quantum Information to Musical Semantics

Maria Luisa Dalla Chiara
Roberto Giuntini
Antonino Riccardo Luciani
Elenora Negri

Illustrated by Cristina Seravalli

ISBN 978-1-904987-61-5

College Publications
Scientific Director: Dov Gabbay
Managing Director: Jane Spurr
Department of Computer Science
King's College London, Strand, London WC2R 2LS, UK

http://www.collegepublications.co.uk

Original cover design by Richard Fraser
Cover produced by Laraine Welch
Printed by Lightning Source, Milton Keynes, UK

Dedicated to the memory of Giuliano Toraldo di Francia,
who has inspired many ideas discussed in this book

Contents

Introduction

Albert Einstein has once observed that there is *something musical* in the structure of quantum mechanics. What did the great physicist, who liked to play violin, really mean by this somewhat mysterious remark?

Einstein's attitude towards quantum theory (QT) has been, in a sense, paradoxical. For, he has been at the same time, one of the "founding fathers" and one of the most serious critics of the theory. While his idea about the *musicality* of QT is not well known, one often quotes his claim "I cannot believe that God is playing dice".

A rigorous criticism of QT has been developed in a celebrated article, jointly written with Podolsky and Rosen, "Can quantum mechanical description of reality be considered complete?" This is the paper where one of the most famous quantum theoretic paradoxes (in jargon termed "EPR-paradox") has been presented for the first time. There is something paradoxical in the history of this paradox. Most of the features that Einstein Podolsky and Rosen had described as negative consequences of QT have, later on, been converted into theoretic and practical advantages, even from the technological point of view. We need only think that *teleportation*, *quantum computation* and *quantum cryptography* systematically use, in a positive way, some characteristic quantum phenomena that are usually called *EPR-situations*.

Beyond the intentions of the authors, the EPR-article has stimulated a genuine "epistemological revolution", forcing us to discuss, according to new modalities, some crucial problems of the theory of knowledge: the relationship between an *observing subject* and an *observed object*, the notions of *physical object* and of *physical property*, the relation *whole-parts* for physical objects, the *space-time framework* and the role of *probability*.

We are now beginning to understand how the mysterious *quantum entanglement* (which represents the most intriguing feature of the EPR-situations) can be also applied to a formal analysis of some characteristic semantic phenomena, where *holistic* and *contextual* aspects play a relevant role.

As is well known, the traditional semantic theories, based on classical logic, are "desperately" *analytical* and *antiholistic*. A basic principle in these theories is a *compositionality-assumption*, according to which the meaning of any compound expression is determined by the meanings of its parts. Furthermore, meanings are always supposed to be *precise* and *non-ambiguous*. As a consequence, classical semantics turns out to hardly applicable to an adequate formal analysis either of natural languages or of the languages of art, where contextuality and ambiguity seem to represent essential features. The quantum-theoretic formalism, instead, gives rise to some characteristic *entangled states of knowledge* where our information about the *whole* determines our information about the *parts*. And the procedure cannot be, generally, inverted: in other words, it is impossible to reconstruct the global information as a mere combination of the partial pieces of information about the component elements. Once broken into its parts, the puzzle cannot be composed again!

To what extent is it reasonable and interesting to apply some general ideas that arise in the "far" quantum world in order to reconstruct formally the deep semantic structures that underly musical compositions? In this book, we will try and give some answers to this question.

CHAPTER 1

Quantum enigmas

Quantum theory (QT) investigates the behaviour of "very small" physical objects (photons, electrons, protons, quarks, ...). Is it sensible to extend such a theory beyond its natural borderlines, trying to apply the quantum theoretic concepts to a "very far" field like the world of music? The risk is violating one of the basic principles of scientific method, according to which any scientific theory has a characteristic *validity-domain*. Accordingly, all extrapolation-attempts (which transfer concepts and results across different domains) may be, in principle, dangerous and misleading. However, things are not so simple. In fact, the "strange" behaviour of microobjects (which seem to be completely different from the objects we are accustomed to meet in our common life experience) has represented a kind of "alarm-bell" for the scientific community, forcing us to discuss some basic concepts of the theory of knowledge according to new modalities. After the "quantum revolution" one cannot avoid reconsidering, in a new light, some questions that had been traditionally investigated by scientists and philosophers:

- what exactly are *objects*, *properties* and *relations*?
- What is the role played by *observers* with respect to *observed objects*?
- What is the role of *probability*?
- Does the notion of *cause* still have a scientific legitimacy?
- What do *time-evolution* and *history of a given individual* exactly mean?

Apparently, the partial answers given by QT to such questions have a bearing even outside the universe of microobjects. We need only think that *uncertainty*, *indetermination* and *ambiguity* seem to be characteristic features of quantum phenomena and of human situations, as well. On this basis, one can reasonably propose the following conjecture: there is something "universal" in the quantum theoretic formalism, which might be successfully applied even to fields that are very far from microphysics.

One can recognize a deep coherence between the basic ideas of classical physics and of classical logic. In both cases, our representation of the *external world* and of our *thoughts about the external world* seems be grounded on some general principles that can be summarized as follows:

- any theory refers to some *universes of discourse*, consisting of *objects* (or *individuals*) that satisfy some *properties* and *relations*.

- All problems expressed in the language of a theory are *semantically decided*: according to the (semantic) *excluded-middle principle* any sentence of the language is either *true* or *false* (*tertium non datur!*). Of course, the principle does not imply that the scientific community is able to *solve* all the problems that are currently investigated. For a hypothetical omniscient mind, however, any question that is rationally formulated admits either a positive or a negative answer.
- Contradictions are strongly forbidden in a double sense. By the *semantic non-contradiction principle*, a sentence A and its negation $\neg A$ cannot be true at the same time. Furthermore, by the *logical non-contradiction principle*, any contradictory sentence $A \wedge \neg A$ (A *and not* A) is always false.
- All meanings of scientific expressions should be, in principle, precise and non-ambiguous. As a consequence, fuzzy or indeterminate semantic situations are not admitted.
- Semantic rules are supposed to have a *compositional* behaviour: the *meaning* of any compound expression (say, "2 is even and 3 is odd") is generally determined by the meanings of its parts (subexpressions).

We will see how all these principles (which seem to be *prima facie* very natural) have been put in question by the quantum world.

The origin of all logical anomalies of QT can be recognized in a general principle of the theory that has sometimes been described as somewhat "scandalous". This is the *superposition principle* according to which

pure states can be summed, giving rise to new pure states.

Why is such a statement strange and what are exactly *pure states*? As is well known, any physical theory (either in the classical or in the quantum case) describes domains of physical objects, by associating to the objects under investigation possible *states* that sum up the observer's knowledge. In the happiest situations, a state s represents a *maximum of information* that cannot be consistently extended to a richer knowledge (expressed in the language of the theory). Even a hypothetical omniscient mind could not know more about the object in question! Non maximal pieces of information, instead, are represented by *mixtures of pure states* (also called *mixed states*). Intuitively, a mixed state can be thought of as a particular ensemble of pure states, each of them having a *weight* in the interval $[0, 1]$. At first sight, physical pure states seem to be close to the notion of *complete concept* of a given individual, discussed by Leibniz: although human knowledge is unavoidably partial and limited, God knows the complete concepts of all objects living either in the real or in some possible world. Like the pure states of physics, Leibniz' complete concepts represent maximal and consistent pieces of information about the corresponding objects: adding a new information (expressed in the same language) would determine a collapse of the complete concept into a contradictory description. Metaphorically, complete concepts can be compared to maximally blown up balloons: trying to inflate them a little bit more would let them explode.

Both in classical logic and in classical physics, maximal pieces of information turn out to satisfy an important property, which seems to be quite natural from an intuitive point of view. Whoever has a maximum of information about a given object is always able to *decide* all properties and relations that concern the object in question. Although many properties of the Moon are unknown to human persons, an omniscient mind, grasping the Moon's complete concept, would be able to give a positive or a negative answer to any reasonable question, formulated in the language that describes the Moon's world. Accordingly, we can say that any maximal information about a given system of objects is *logically complete*, in the following sense: for any sentence A (expressed in the language of the system), our information decides whether A or its negation $\neg A$ holds.

Asserting that any maximal information is logically complete seems to be, *prima facie*, quite natural. Suppose, for instance, that the Moon's complete concept does not decide a given sentence A (say, "there is water on the Moon"). In such a case, the concept in question cannot represent a maximal information. In fact, we might add to our available information either A or $\neg A$, obtaining in this way a new consistent and richer information. Hence, one could inflate the balloon, without letting it explode! Strangely enough, this argument, that may appear logically universal, turns out to be violated in the quantum world. How is such a "logical deviation" possible?

By Heisenberg's *uncertainty principle* (one of the basic tenets of QT), quantum objects are essentially "poor": the "poverty" regards the number of properties that can be satisfied at the same time. One is dealing with a kind of limitation that is deeply different from the limitations we are accustomed to meet in the classical world. Of course, nobody expects that electrons or photons satisfy properties that have no physical sense for them (say, "being hot" or "being coloured"). However, what may appear strange is the following phenomenon: some properties (which are perfectly defined for quantum objects) can be semantically decided only in particular situations. In some other situations these same properties are strongly undecidable even for a hypothetical omniscient mind. In fact, according to Heisenberg's principle, there are pairs of *incompatible* physical quantities for which a quantum object cannot have a precise value at the same time. Suppose, for instance, that a particle has a precise velocity-value; then, its position shall be completely indeterminate. And viceversa. By using a joky metaphor, we might say that giving a fine for excessive speed is impossible in the quantum world! A quantum policeman, knowing the exact speed of a given electron, would not be able to capture it, because our electron would be, in principle, completely delocalized. Viceversa, should the policeman succeed in capturing the electron (determining its exact position), he would not be entitled to give a fine, because, in such a case, the speed-value would

FIGURE 1. The quantum policeman

be completely indeterminate (Fig.1). All this brings about an immediate consequence for the logical completeness question. Any maximal information (i.e. any pure state) that assigns a precise velocity-value to a quantum object shall leave completely indeterminate all the properties that concern precise position-values. And viceversa. Therefore, all quantum pure states are examples of maximal pieces of information that are logically incomplete.

Now we can understand how the strange *superpositions of pure states* emerge in QT. In the classical case, trying to make a *sum* (or a *union*) of different maximal pieces of information would not have any sense. For instance, putting together two different complete concepts of the Moon would certainly lead to a contradictory description (our balloon would explode!). We can easily see why. Suppose that $\mathcal{C}_1$ and $\mathcal{C}_2$ represent two different complete concepts of the Moon. Just because they are different, there will be at least one sentence A (say, "there is water on the Moon") such that $\mathcal{C}_1$ asserts A, while $\mathcal{C}_2$ does not assert A. We know, however, that classical complete concepts are logically complete. The concept $\mathcal{C}_2$ that does not assert A, shall assert $\neg A$ ("there is no water on the Moon"). As a consequence, $\mathcal{C}_1$ and $\mathcal{C}_2$ are contradictory; hence their "sum" $\mathcal{C}_1 + \mathcal{C}_2$ represents an impossible description. There is no complete concept $\mathcal{C}$ such that $\mathcal{C} = \mathcal{C}_1 + \mathcal{C}_2$!

What happens in the quantum case is fairly different, because quantum pure states represent pieces of information that are maximal and logically incomplete at the same time. Suppose that $\mathcal{C}$, $\mathcal{C}_1$, $\mathcal{C}_2$ correspond to three possible states of a quantum object (say, an electron). What might be the meaning of a sum having the form $\mathcal{C} = \mathcal{C}_1 + \mathcal{C}_2$? The basic idea is the following: such a sum represents a kind of "interlacement" between *real* and *potential* properties of the object described by $\mathcal{C}$. We know that $\mathcal{C}$ only decides a few properties of the object in question: a number of relevant properties turn out to be indeterminate. The states $\mathcal{C}_1$ and $\mathcal{C}_2$ have just the role of describing the "cloud of the potential properties" of our object. In a sense we can say: the actual state of affair is described by $\mathcal{C}$; however, things *might* also be the case according to $\mathcal{C}_1$ or according to $\mathcal{C}_2$.

How are pure state represented in the framework of the mathematical formalism of QT? We should not imagine a state as a piece of information expressed by means of linguistic expressions. In fact, states are identified with some special mathematical objects, "living" in some abstract spaces that represent the *mathematical environment* for the physical objects under investigation. Technically, such spaces are called *Hilbert spaces*: one is dealing with a particular kind of vector spaces, which represent a generalization of the Cartesian plane and of Euclidean spaces. In such a context, pure states are identified with unit vectors (whose length is 1) of a convenient Hilbert space $\mathcal{H}$; following a happy notation (introduced by Paul Dirac), they are usually indicated by the symbols $|\psi\rangle$, $|\varphi\rangle$, $|\chi\rangle$, ... On this basis, the axioms of QT permit us to prove that any pure state $|\psi\rangle$ assigns a precise probability-value to any physical property asserting that a given quantity has a certain value. For instance, a state $|\psi\rangle$ might assign probability 1 to the property

the velocity-value is c (the light-speed).

In this way, any state $|\psi\rangle$ turns out to describe, at the same time, all *certain* and all *potential* properties of the object under investigation, by associating a precise probability-value to each possible property.

We can now abandon our rough way of indicating sums of maximal pieces of information. Following the usual notation of QT, instead of $\mathcal{C} = \mathcal{C}_1 + \mathcal{C}_2$, we will write

$$|\psi\rangle = c_1|\psi_1\rangle + c_2|\psi_2\rangle,$$

where $|\psi\rangle$, $|\psi_1\rangle$, $|\psi_2\rangle$ represent three possible pure states, while c_1 and c_2 (also called *amplitudes*) are two complex numbers, whose role is determining two special probability-values. The intuitive interpretation of such a sum can be described as follows:

> **the object under investigation certainly satisfies all properties to which $|\psi\rangle$ assigns probability 1. At the same time, our object might satisfy with probability $|\, c_1 \,|^2$ all properties that are certain for state $|\psi_1\rangle$,**

and might satisfy with probability $\mid c_2 \mid^2$ all properties that are certain for state $|\psi_2\rangle$.[1]

In such a case the state $|\psi\rangle$ is called a *superposition* of the states $|\psi_1\rangle$ and $|\psi_2\rangle$. As expected, superpositions do not necessarily consist of two elements, only. Generally, we will have finite and also infinite superpositions, which will be indicated as follows:

$$|\psi\rangle = c_1|\psi_1\rangle + \ldots + c_n|\psi_n\rangle \text{ and } |\psi\rangle = \sum_{i\in\mathbb{N}} c_i|\psi_i\rangle.$$

The superposition-condition of all quantum objects has some critical consequences that concern the interaction between an *observer* and an *observed object* during a measurement procedure. As is well known, in classical physics any measurement-performance represents a typical *epistemic discovery*. Whenever we put a given object on a balance in order to test its weight, we do not doubt that the weight-value pre-existed to our action. Of course, we are aware that, in some particular circumstances, the interaction with a measurement apparatus might slightly transform the value of the measured quantity. However, in such a case, nobody believes that the apparatus itself has "created" the measurement-result!

In QT measurement-procedures give rise to more problematic situations. Consider an electron,whose pure state has the following form:

$$|\psi\rangle = c_1|\psi_1\rangle + c_2|\psi_2\rangle.$$

Suppose that such superposition describes the following physical situation (Fig.2):

- state $|\psi\rangle$ does not determine the position of the electron, which is delocalized;
- state $|\psi_1\rangle$ decides that the electron is outside a given box;
- state $|\psi_2\rangle$ decides that the electron is inside the box.

Suppose now that an observer, whose initial information corresponds to state $|\psi\rangle$ decides to test the electron's position, discovering that the electron is inside the box. As a consequence of the acquired information, our observer will naturally abandon his (her) initial information, assuming the new information $|\psi_2\rangle$, which decides that the electron is inside the box. This determines a *state-transformation*, which can be schematically represented as follows:

$$|\psi\rangle \mapsto |\psi_2\rangle.$$

Thus, after the measurement, state $|\psi\rangle$, for which the property "being inside the box" was indeterminate, is transformed into state $|\psi_2\rangle$, which decides the property in question. One might have the impression that our

[1]For any complex number z, $\mid z \mid$ represents the non-negative real number that is the *modulus* of z. Since the length of $|\psi\rangle$ is 1, the numbers $\mid c_1 \mid^2$ and $\mid c_2 \mid^2$ turn out to behave as convenient probability-values (belonging to the interval $[0, 1,])$.

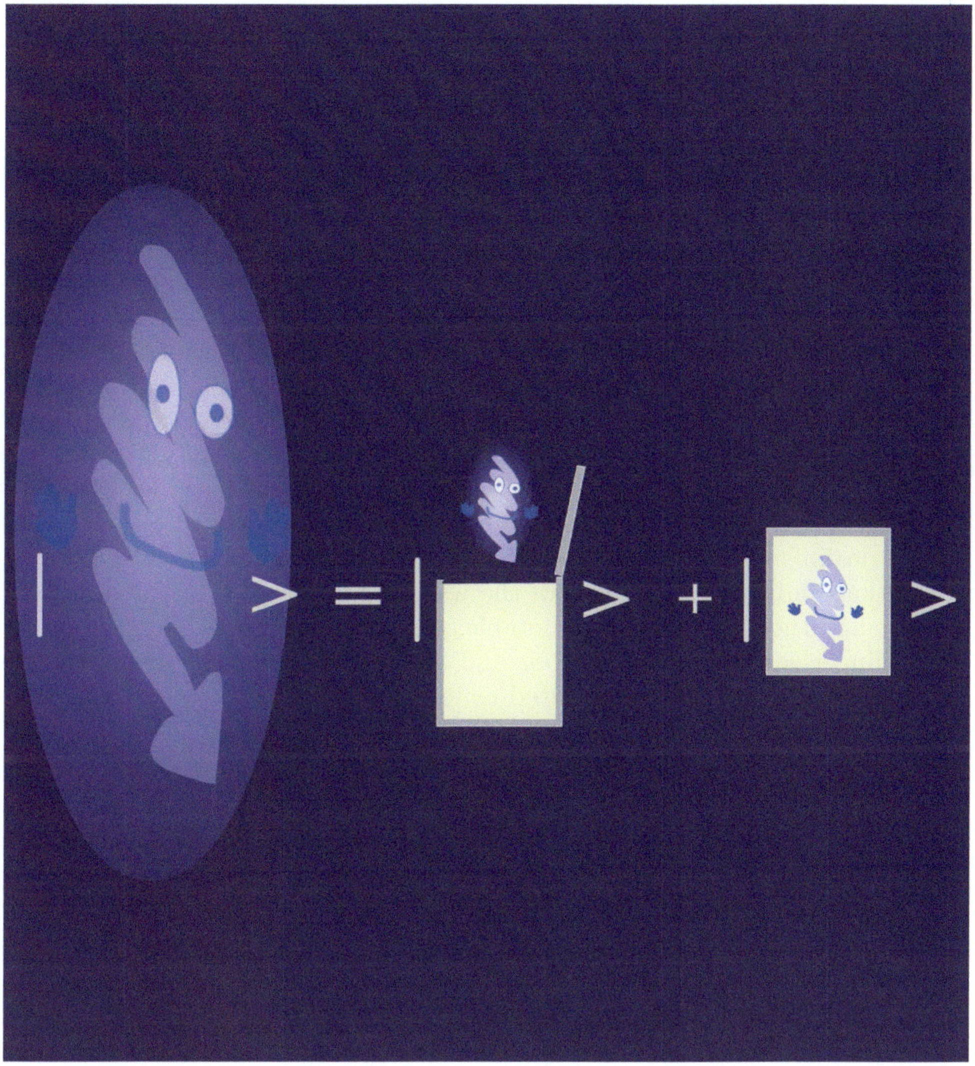

FIGURE 2. The electron might be either in the box or outside.

electron (which, before the measurement, had a kind of "ghost-behaviour") has been "materialized" within a precise space-region. In such cases, according to the standard quantum theoretic jargon, one uses to say that the state-transformation has been performed by *collapse* (or *reduction*) *of the wave-function*.

The *collapse-principle* has been proposed for the first time by John von Neumann as one of the basic axioms of QT.[2] At a first reading, the principle

[2]See J. von Neumann, *Mathematische Grundlagen der Quantenmechanik*, Springer, Berlin, 1932. English translation: *Mathematical Foundations of Quantum Theory*, Princeton University Press, Princeton, 1996.

might appear somehow trivial: whenever we acquire a new information by means of a measurement-procedure, we naturally take into account the obtained result: consequently, we transform our initial knowledge into a new one. However, in the quantum case, things are not so simple. Generally, the observer does not discover something that pre-existed to his (her) act of measuring. In fact, state $|\psi\rangle$ does not represent our *ignorance*, but rather an *objective indeterminacy* about the electron's localization. Only the choice of the observer, who decides to measure position instead of other possible quantities, brings into being something that was before strongly indeterminate, even *in mente Dei*. One might believe that the act of observing has a kind of "magic power", which transforms potential properties into actual ones.

What happens in a situation where, instead of performing a measurement, we let our electron evolve spontaneously? In such a case, another basic axiom of the theory (the *Schrödinger-equation*) determines the state-evolution of an object that is not interacting with any observer or apparatus. One is dealing with a completely different state-transformation. While transformations induced by collapse of the wave-function are discontinuous, stochastic and theoretically unpredictable, the time-evolution described by the Schrödinger-equation is *continuous* and *linear*. Schematically, Schrödinger's transformations can be represented as follows:

$$|\psi(t_0)\rangle \quad \mapsto |\psi(t_1)\rangle = U[t_0, t_1](|\psi(t_0)\rangle),$$

where $|\psi(t_0)\rangle$ and $|\psi(t_1)\rangle$ represent the states of a given object at time t_0 and t_1, respectively, while $U[t_0, t_1]$ is a *unitary* operator that describes the time-evolution of our object during the time-interval $[t_0, t_1]$.[3]

As a consequence of the linearity of $U[t_0, t_1]$, the Schrödinger-evolution cannot resolve any initial uncertainty. Suppose that the state of a given object at the initial time t_0 is the superposition

$$|\psi(t_0)\rangle = c_1|\psi_1\rangle + \ldots + c_n|\psi_n\rangle),$$

then the state of the object at the final time t_1 will be

$$|\psi(t_1)\rangle = c_1 U[t_0, t_1](|\psi_1\rangle) + \ldots + c_n U[t_0, t_1](|\psi_n\rangle),$$

which preserves the same uncertainty-degree of the initial state.

One can easily understand how the co-existence in one and the same theory of two different axioms like von Neumann's collapse of the wave-function and Schrödinger's equation may give rise to a conflict-situation. We will discuss this problem in the following Chapter, where we will illustrate the Einstein-Podolsky-Rosen paradox.

---

[3]Unitary operators U have two important properties: 1) they preserve the vector-length (hence pure states are transformed into pure states); 2) being *linear* operators, they preserve the superposition-form (hence $U(c_1|\psi_1\rangle + \ldots + c_n|\psi_n\rangle) = c_1 U(|\psi_1\rangle) + \ldots + c_n U(|\psi_n\rangle)$).

CHAPTER 2

The EPR-paradox and Susan's "true history"

1. Remote actions

We will now try and illustrate the EPR-paradox[1] in a simple way, using a metaphorical description (which will preserve the logical structure of the argument). The paradox refers to a particular physical situation that involves two quantum particles (say, two electrons), which will be called *Sarah* and *Susan* in the framework of our metaphor. We suppose that the two objects have interacted in the past and are separated after a given time t_0. One is dealing with a "very strong" (*space-like*) separation: in other words, *Sarah* and *Susan* cannot exchange any signal during the time-interval we are referring to. Two observers *Oswald* and *Oscar* are observing *Sarah* and *Susan*, respectively. We are interested in a pair of *incompatible physical quantities*, indicated by $\mathcal{P}$ and $\mathcal{Q}$, that cannot be simultaneously measured, by Heisenberg's *uncertainty principle* (for instance, the quantity *spin* in the x direction and the quantity *spin* in the y direction). In our metaphor, we can imagine that $\mathcal{P}$ represents the hair-colour, while $\mathcal{Q}$ represents the eyes-colour. Both quantities $\mathcal{P}$ and $\mathcal{Q}$ can only assume two possible values: + (positive) and − (negative). As a consequence, we obtain two pairs of possible properties

$$(P^+, P^-) \text{ and } (Q^+, Q^-)$$

(say, (*dark hair*, *light hair*) and (*dark eyes*, *light eyes*)).

Owing to the past interaction, there is a strong correlation between *Sarah* and *Susan*, which concerns the values of the quantities under consideration:

Sarah (Susan) verifies P^+ [*has dark hair*]
if and only if
Susan (Sarah) verifies P^- [*has light hair*].

A similar relation holds for the pair Q^+, Q^- (dark eyes/light eyes).

Suppose that at the initial time t_0 the two quantities $\mathcal{P}$ and $\mathcal{Q}$ are totally indeterminate both for *Sarah* and for *Susan*. Soon before time t_2 (in the interval $[t_1, t_2]$) *Oswald* decides to measure quantity $\mathcal{P}$, discovering that

Sarah is P^+ [*has dark hair*].

In the metaphor, *Oswald* might take *Sarah*'s hat off, discovering that she has dark hair (Fig. 1).

[1] A. Einstein, B. Podolsky and N. Rosen, "Can quantum-mechanical description of reality be considered complete?", *Physical Review* **47**, pp. 777–780, 1935.

FIGURE 1. Sarah's and Susan's histories

Hence, owing to the correlation between *Sarah* and *Susan*, *Oswald* knows, without having in any way interacted with *Susan*, that

Susan is P^- [*has light hair*].

At the same time, we suppose that the other observer *Oscar* did not perform any measurement (he did not take either *Susan*'s hat or *Susan*'s glasses off). A question arises: how to determine *Sarah*'s and *Susan*'s "true histories" during the time-interval $[t_0, t_2]$? For *Sarah*, the most natural answer seems to be the following: her history is determined by the state-sequence

$$(s^a(t_0), s^a(t_1), s^a(t_2)),$$

where $s^a(t_1)$ is obtained from $s^a(t_0)$ by application of the Schrödinger-equation, while $s^a(t_2)$ is obtained by collapse of the wave-function. Property P^+, which is indeterminate for states $s^a(t_0)$ and $s^a(t_1)$, is decided for state $s^a(t_2)$ (because *Oswald* has discovered that *Sarah* has dark hair). What about *Susan*? At first sight, two different histories seem to be legitimate. The first one is the history that is "seen" by *Oscar*, who did not perform any measurement:

$$(s^u(t_0), s^u(t_1), s^u(t_2)),$$

where all states are determined by the Schrödinger-equation and where at any time properties P^+ and P^- are undecided. The second one is the history that is "imagined" by *Oswald*, who (having measured quantity $\mathcal{P}$ on *Sarah*) knows that at time t_2

Susan is P^- *[has light hair].*

As a consequence, the history imagined by *Oswald* will be the following:

$$(s^u(t_0), s^u(t_1), s^{u^*}(t_2)),$$

where the final state $s^{u^*}(t_2)$, obtained by collapse of the wave-function, decides that *Susan* is P^- (has light hair). Is it reasonable to ask (as we sometimes do in real life): which is the "true history" of *Susan*?

The concept of "true history" turns out to be deeply connected with the critical *reality-principle*, which represents the main philosophical hypothesis of the EPR-argument. The principle is formulated as follows:

> *If, without in any way disturbing the system, we can predict with certainty (i.e., with probability equal to unity) the value of a physical quantity, then there exists an element of reality corresponding to that physical quantity.*

Apparently, the reality-principle proposes a *sufficient condition* for a physical property to be considered *objective* (i.e. independent of any observer's action). Let us ask: does P^- (having light hair) represent an objective property for *Susan*? A positive answer to this question seems to be justified by the reality-principle combined with a *locality hypothesis*, according to which *no superluminal influence* is admitted. *Susan* cannot be disturbed by any action performed by a far observer like *Oswald* (by the locality-principle). As a consequence, by the reality-principle, P^- represents an *objective property* for *Susan*.

We can now use a *counterfactual argument* (based on an implication whose antecedent is false). During the time interval $[t_1, t_2]$, *Oswald* might have chosen to measure quantity $\mathcal{Q}$ instead of quantity $\mathcal{P}$. As a consequence, the following counterfactual implication must be correct:

Should Oswald have chosen to measure $\mathcal{Q}$*, then either* Q^+ *or* Q^- *would be an objective property for Susan at time* t_2*.*

However, by definition of objectivity, *Susan*'s objective properties cannot be determined by *Oswald*'s choices (which depend on his free will). Hence, we obtain:

Either $[P^-$ *and* $Q^+]$

or

$[P^-$ *and* $Q^-]$

is an objective property for Susan at time t_2.

This conclusion is justified by the objective character of property P^- and by a natural logical principle according to which *the conjunction of two objective properties is an objective property*. At this point, the argument invokes another important hypothesis, the *physical completeness* of QT, formulated by Einstein, Podolsky and Rosen as follows:

> *Every element of physical reality must have a counterpart in the physical theory.*

In other words: if the theory is physically complete, then all objective properties (expressed in the language of the theory) should be reflected by some corresponding pure states.

Hence, in particular, there should be a pure state of *Susan* that assigns probability 1 to one of the two following properties:

1) $[P^-$ *and* $Q^+]$ [*light hair and dark eyes*];
2) $[P^-$ *and* $Q^-]$ [*light hair and light eyes*].

However this result contradicts the uncertainty principle: the two quantities $\mathcal{P}$ and $\mathcal{Q}$ are incompatible; consequently, no state can assign a precise value to both of them. We have obtained, in this way, a formal contradiction that seems to put in question the logical soundness of quantum mechanics.

Is it possible to block the derivation of such a paradox? The proof of a contradiction in a scientific theory is, in a sense, similar to the discovery of a murder in the framework of a detective story (Fig. 2). And each solution that is proposed to avoid the contradiction plays the role of a detective who identifies the murderer. Of course, as happens in detective stories, also scientific paradoxes generally give rise to different possible solutions. In the case of the EPR-argument the possibly "guilty" hypotheses are:

1) the reality-principle;
2) the locality-principle;
3) the physical completeness-principle.

Each solution is characterized by a different choice of some guilty hypotheses. Einstein, Podolsky and Rosen had no doubts: the hypothesis, that has to be refused, is the physical completeness-principle. In fact, the original version of the EPR-argument was presented as a kind of *proof by contradiction* whose conclusion was: *QT is physically incomplete*. In other words, the pure states of the theory do not represent a *maximum of information*; one is dealing with a kind of *statistical pieces of information*, that are quite similar to

FIGURE 2. Proving a contradiction in a theory is like discovering a murder in a detective story

the states of classical statistical mechanics. The article "Can quantum-mechanical description of reality be considered complete?", concludes as follows:

> *While we have thus proved that the wave function does not provide a complete description of the physical reality, we left open the question of whether or not such a description exists. We believe, however, that such a theory is possible.*

However, this conclusion contains a "logical mistake". In fact, the EPR-argument only proves the logical incompatibility between QT and our three general principles, without forcing us to choose a particular "guilty hypothesis". There are other legitimate solutions, which make different choices. For instance, the solution proposed by Niels Bohr and by the "Copenhagen-school" is based on the refusal of the reality-principle. According to Bohr, it is not sensible to speak of "elements of reality", because *all properties of*

FIGURE 3. Quantum non-locality does not imply the possibility of transmitting a signal. Oswald should use a normal physical channel (like a phone) in order to send information to Oscar

physical objects have to be described as relations. In our days, many scholars are in favor of solutions based on a refusal of the locality-principle. In this perspective, the action performed by *Oswald* in the interval $[t_1, t_2]$ turns out to have a "genuine physical influence" on *Susan*, in spite of the space-like separation. However, this result does not imply the possibility of sending a signal from *Oswald* to *Oscar* during the interval $[t_1, t_2]$. Accordingly, there is no conflict between quantum non-locality and special relativity (which was Einstein's basic worry) (Fig. 3).

Is it possible to recover, in this way, the *unicity* of *Susan*'s history? The question does not admit an easy answer.

In 1985, half a century after the appearing of the EPR-article, only Nathan Rosen was still alive. A number of conferences were organized to celebrate the discovery of the famous paradox; and sometimes the "honour-guest" was, of course, Rosen. How did the "third man" of the *terzetto* regard the EPR-argument, fifty years after? The following quotation represents an interesting witnessing:

> *At the time of the writing of the EPR paper I agreed with the belief expressed at the end that a complete theory is possible. Since then fifty years have passed and physics has changed greatly. In recent years doubts have arisen in my mind as to whether a theory will be found in the future that will be*

> *complete according to the criteria of the paper and will be correct in giving agreement with observation... Hence it is hard to believe that a theory will be found that will be complete, based on the criterion of an element of reality, used in the paper. It may also be that in the future physical theories will describe reality in different terms from those to which we are now accustomed. Does this means that the EPR paper is useless? I think not. The paper has led to a great deal of discussion that has helped to clarify the physical concepts. I like to believe that this has contributed, if in a small measure, to the progress of physics.*[2]

2. The interaction between actual and possible objects

The strange behaviour of quantum objects has changed some general ideas about the interaction between *what does exist* and *what might exist*. In the framework of classical semantics, *actual* and *possible objects* were sharply distinguished: the set of all *actual objects* was described as a *classical proper subset* (with no *fuzzy* borders) of the class of all possible objects. Quantum logic (created by Birkhoff and von Neumann as a natural abstraction from the mathematical formalism of quantum theory[3]) is, in this respect, "more liberal": actual existences generally depend on virtual existences! As noticed by the philosopher and logician Willard Van Orman Quine, the traditional concept of *physical object* tends to "evaporate" in modern physics.

Any logical investigation about the *existence problem* is of course concerned with the semantic analysis of the *existential quantifier* ($\exists$), which behaves differently in different logics. When discussing these questions, it is useful to refer to a successful and flexible kind of semantics that has been termed *possible world semantics.* The basic intuitive idea is very simple and goes back to Leibniz. We suppose that the *actual world* (indicated by W) is *correlated* with a number of *alternative situations*, also called *possible worlds.* Of course, the actual world itself represents an example of a possible world. The correlation between worlds is usually called *accessibility relation.* From an intuitive point of view, a world W_1 is accessible to another world W_2, when W_2 represents a *reasonable alternative* for W_1.

In spite of its metaphysical aspect, the possible world semantics admits a quite natural physical interpretation and can be applied to quantum logic. The basic intuitive idea is the following: any possible world represents the knowledge of an observer about a quantum object (say an electron or a photon). Hence it can be mathematically represented as a possible state of the object under investigation.

---

[2] N. Rosen, "Quantum mechanics and reality", in P. Lahti and P. Mittelstaedt (eds), *Symposium on the foundations of modern physics*, World Scientific, Singapore, 1985.

[3] G. Birkhoff and J. von Neumann, "The logic of quantum mechanics", *Annals of Mathematics* **37**, pp. 823-843, 1936.

What about the quantum meaning of the accessibility relation? Mathematically, two pure states $|\psi\rangle$ and $|\varphi\rangle$ are accessible when they are *non-orthogonal* in the abstract space $\mathcal{H}$, where all pure states "live". This corresponds to the following physical relation: $|\psi\rangle$ is accessible to $|\varphi\rangle$ when $|\psi\rangle$ can be transformed into $|\varphi\rangle$ after the performance of a measurement concerning a *physical quantity* that can be measured on the system. The state-transformation $|\psi\rangle \mapsto |\varphi\rangle$ (induced by the measurement) is determined by *collapse of the wave-function.*

In order to understand the physical interpretation of the quantum-logical existential quantifier, let us first recall that in most logics the existential quantifier ($\exists$) is a generalization of the disjunction ($\vee$). As an example, let us think of the original human universe represented by the set whose only elements are *Adam* and *Eve*. In this universe, the existential sentence

There exists somebody who ate the apple

is clearly equivalent to the disjunctive sentence

Either Adam or Eve ate the apple.

Of course, in the case of our present human universe, using disjunctive sentences (with about 6 billions of members) instead of existential sentences would be quite unpractical!

Also in quantum logic, the existential quantifier $\exists_q$ is a generalization of the quantum-logical disjunction $\vee_q$. The truth-condition for disjunctive sentences is the following:

Either A ***or*** B

is *TRUE* in the actual world W

$(\models_W A \vee_q B)$

if and only if

for any possible world W_1 accessible to W there exists a possible world W_2 accessible to W_1 such that either A is true in W_2 or B is true in W_2.

Apparently, truth and falsity in the actual world essentially depend on what happens in other possible worlds. As a consequence, a disjunction may be *true* (in the actual world) even if both members are *indeterminate.* Such a semantic behaviour, which may appear *prima facie* somewhat strange, seems to reflect pretty well a number of concrete quantum situations. In fact, in quantum theory one is often dealing with alternatives that are semantically determined and true, while both members are, in principle, indeterminate.

3. The two-slit experiment

One of the fundamental experiments of quantum mechanics, the *two-slit experiment*, has been often described as a kind of *experimentum crucis* for quantum logic. One is dealing with an intriguing result that Richard Feynman (in his celebrated *Lectures on Physics*[4]) has commented as follows:

> *It is all quite mysterious. And the more you look at it, the more mysterious it seems.*

In order to understand, from an intuitive point of view, why the two-slit experiment appears so "mysterious", we will follow an amusing description that can be found in the book *Alice in Quantum Land. An Allegory of Quantum Physics* by Robert Gilmore.[5] During her fantastic travel through the mysteries of the quantum world, Alice arrives at the *Gedanken room*, where all *thought-experiments* become *real*. First the *Classical Mechanic* illustrates the two-slit experiment, using a bullet-gun. His "classical" gun begins firing a number of bullets. Most of them whine off in all directions; but some of them go through a screen with two slits (say slit 1 and slit 2) and finally hit the wall (in front of the screen), each leaving one precise mark. The final configuration of the marks on the wall gives rise to a typical *additive pattern*, representing the *probability* that a generic bullet reaches a given region of the wall (Fig. 4).

The *Classical Mechanic* explains to Alice:

> "Since both slits are open, the bullets may pass through either slit, so the overall distribution is given by the sum of the probabilities that we got for the two slits on their own, since the bullets must have passed through one or the other. They cannot have passed through both, you know."

Soon after, the *Quantum Mechanic* makes a similar experiment, using an *electron gun*. The effect is completely different: a clear *interference pattern* appears, quite similar to what would be the case in a wave-phenomenon (Fig. 5).

The *Quantum Mechanic* explains:

> "There you see a clear *interference effect*. With one slit alone we would have seen that the distribution would decrease smoothly to either side, much as the bullets did. In this case, you see that, when there are two slits open, the *amplitudes* from the two slits are interfering and are producing obvious peaks and troughs in the probability distribution. The behaviour of the electrons is quite different from that of my friend's bullets."
>
> "I do not understand - says Alice - Do you mean that there are so many electrons going through that somehow the electrons which go through one hole are interfering with the ones which go through the other?"

---

[4]R. P. Feynman, R. B. Leighton, M. Sands, *The Feynman Lectures on Physics*, Addison Wesley, London, 1969.

[5]R. Gilmore, *Alice in Quantum Land. An Allegory of Quantum Physics*, Sigma Press, Wilmslow, 1994.

FIGURE 4. Alice, the Classical Mechanic and the two-slit experiment

> "No. Not at all - replies the *Quantum Mechanic* - The interference effect works even when there is only one electron present in any time. One electron on its own can show interference. It can go through both slits and interfere with itself."

> "But that is silly! - cries Alice - One electron cannot go through both slits. As the *Classical Mechanic* said, it just isn't sensible."

Quickly she directs the light toward the two slits:

> "I can see the electrons as they go through the slits and it is just as I said it must be. Each one does go through just one slit."

FIGURE 5. Alice, the Quantum Mechanic and the two-slit experiment

> "Aha! - replies the *Quantum Mechanic* - but have you looked to see what is happening to the interference pattern? The interference effects only happen when there is no way that you could know which slit the electron went through. Whether or not you do know does not matter. So you see, when there is interference it seems as if each electron is going through both slits. If you try and check on this you will find that the electrons go through one slit, but then the interference vanishes. You can't win!"(Fig. 6)

Apparently, the case of the interference pattern, arising when no observation is possible, gives rise to a typical quantum-logical behaviour of the connective *or*. The disjunction:

Either the electron has gone through slit 1

or

the electron has gone through slit 2

FIGURE 6. When Alice observes the interference disappears

is true. At the same time, both members of the disjunction are indeterminate and consequently not true. When, instead, *observing* is possible (even if no *actual observer is observing*), the interference pattern disappears and the connective *or* behaves according to classical logic. The truth of the disjunction

Either the electron has gone through slit 1

or

the electron has gone through slit 2

implies that each electron has gone through a well determined slit.

As a conclusion, two different logics seem to co-exist in quantum theory. Microobjects behave according to classical logic in some situations; at the same time, in other situations, they follow the rules of quantum logic. One is dealing with a kind of "logical schizophrenia" that turns out to be somewhat mysterious.

CHAPTER 3

Quantum computers, Gestalt psychology and semantics

Situations that are quite similar to what happens in the two-slit experiment arise in the framework of quantum computation: the strong *parallelism*, characteristic of quantum computers, is essentially based on *virtual paths*, that a quantum computer can follow at the same time. *Reality is, in a sense, entangled with possibility.*

Using quantum indeterminism in order to construct powerful and efficient computers is a brilliant idea that has been first proposed by Richard Feynman.[1] His basic intuition, which might appear, *prima facie*, a kind of ingenious trick, can be sketched as follows. Suppose we want to calculate by means of a traditional computer the time-evolution of a quantum system (which has a finite number of *degrees of freedom*). Generally, such a computation will be extremely complicated, possibly requiring a time that might be longer than the Universe's life itself. Feynman's tricky idea is that we could try and utilize this difficulty in a positive way, by inverting the relationship between the quantum system and the computer. The fundamental hypothesis is the following: it is just the time-evolution of our quantum system that *simulates* a very complex computation. On this basis, the notion of *quantum computer* is identified with a special kind of quantum object that, evolving in time, executes a given computation. Hence, *reading* the computation-result means *performing a quantum measurement*, by applying the *collapse of the wave-function*. What is important from a practical point of view is that such a process can be realized during a relatively "short" time, accessible to the scientific community. As a consequence the notion of *quantum computation* turns out to be deeply different from classical computations and from classical logical proofs. Since it is, in principle, impossible to control step by step what happens "inside the computer-box", we should trust the spontaneous time-evolution of our machine, which is guaranteed by the correct quantum-theoretic laws. In this way (in contrast with traditional logical conceptions), computations and proofs are no longer describable as *pure formal concepts*: for, they are essentially dependent both on the laws of physics and on technological developments.

[1]See R. P. Feynman, "Simulating physics with computers", *International Journal of Theoretical Physics* **21**, pp. 467–488, 1982 and R. P. Feynman, "Quantum mechanical computers", *Optics News* **11**, pp. 11–20, 1985.

Quantum computation has recently suggested some new forms of quantum logic (called *quantum computational logics*), where *meanings* of sentences are identified with quantum information quantities. This provides a mathematical formalism for an abstract *theory of meanings* that can be applied to investigate different kinds of semantic phenomena where *holistic*, *contextual* and *gestaltic* patterns play an essential role (from natural languages to musical compositions).

As is well known, human perception like thinking seems to be essentially *synthetic*. We never perceive an object by *scanning* it point by point. We instead form right away a *Gestalt*, i.e. a global idea of it. Rational activity, as well, seems to be essentially based on *gestaltic patterns*. As an example, we might refer to a case that concerns the chess game. Some time ago, the powerful computer *Deep Blue* has beaten the world champion Gary Kasparov (Fig. 1).[2]

One tells that Kasparov protested, suspecting that some human intervention acted behind the computer. Why? As is well known, the total number of possible games is finite. As a consequence, the games could be divided into three sets:

1. the set of all games won by *white*;
2. the set of all games won by *black*;
3. the set of all games ending in a draw.

Why could not a fantastic computer of the white player always choose from the first set? A strong player certainly must perform some rapid calculations, but above all he must be able first to perceive a *Gestalt* of the position and then to assess by experience the probability of its different issues. Did Kasparov have a hunch of this sort?

Gestalt-thinking cannot be adequately represented in the framework of classical semantics, which is basically *analytical* and *compositional*: the meaning of a *compound* expression is always determined by the meanings of its *parts*. At the same time, meanings are *non-ambiguous* and *sharp*.

All this renders classical semantics hardly applicable to an adequate analysis of natural languages and of artistic contexts, where holistic and ambiguous features seem to play a relevant role. As a significant example, we might refer to the final verse of the poem *L'Infinito* by Giacomo Leopardi:

E 'l naufragar m'è dolce in questo mare
(And drowning in this sea is sweet to me).

This verse has been compared with the last words of Isolde in Wagner's *Tristan und Isolde*:

ertrinken, versinken, unbewusst, höchste Lust![3]

---

[2]See G. Toraldo di Francia, "Deep Blue and the strategy of human thought", *Foundations of Science* **4**, pp. 83–93, 1999.

[3]Drowning, sinking, unconsciously, highest bliss!

FIGURE 1. A challenge between the champion Kasparov and the computer *Deep Blue*

In Leopardi (but, in a similar way, also in Wagner), the poetic result seems to be essentially connected with the following semantic relation: the meanings of the component expressions "naufragar" (drowning), "dolce" (sweet), "mare" (sea) do not correspond here to the most common meanings. By the way, there is no sea in Recanati, Leopardi's native village which the poem refers to. However the usual meanings of our expressions are somehow present and ambiguously correlated with the metaphorical meanings that are evoked by the whole poem. Needless to say, this represents a quite typical semantic situation in poetry (Fig. 2).

In the semantics of quantum computational logics the following conditions are satisfied:

1) *global meanings* (which may correspond to a *Gestalt*) are intrinsically *vague*, because they leave semantically undecided many relevant properties of the objects under investigation;

2) any global meaning determines some *partial meanings*, which are generally vaguer than the global one;

FIGURE 2. How an artificial intelligence might interpret the last verse of Leopardi's *Infinity* according to a compositional semantics

3) meanings (*Gestalten*) can be generally represented as *superpositions* of other meanings, possibly associated to probability-values;

4) meanings (in the same way as *Gestalten*), are dealt with as intrinsically dynamic objects.

In this framework, the *meaning* of any sentence is identified with a quantum information quantity: a system of *qubits* or, more generally, a *mixture* of systems of qubits (also called *qumix*).

1. Bits and qubits

What is a qubit? From an intuitive point of view, a *qubit* can be regarded as the quantum variant of the classical notion of *bit*. As is well known, in classical information theory, one bit measures the information that is transmitted (or received), whenever one chooses one element from a set consisting of two elements (for instance, from the set consisting of the answer *YES* and of the answer *NO*, or from the set consisting of the number 1 and of the number 0). In quantum information, one cannot generally refer to precise answers (like *YES* or *NO*). The typical ambiguous answer is represented by a *quantum perhaps*, that can be described as a quantum *superposition* of the answer *YES* and of the answer *NO*.

From the physical point of view, a qubit can be regarded as the *pure state* of a single particle, while a system of n qubits (also called *quregister*) corresponds to the state of a composite system consisting of n particles. The idea is that a single particle (like an electron) can physically carry the information-quantity represented by one qubit. In order to carry the information stored by n qubits we need, of course, a composite system consisting of n particles.

From the mathematical point of view, qubits are vectors of length 1 that "live" in the two-dimensional Hilbert space $\mathbb{C}^2$ (based on the set of all ordered pairs of complex numbers). Hence, the mathematical form of a qubit is usually written as follows:

$$|\psi\rangle = a|0\rangle + b|1\rangle,$$

where $|0\rangle = (1,0)$ and $|1\rangle = (0,1)$ are the two elements of the canonical orthonormal basis of the space.[4] From an intuitive point of view, the vector $|0\rangle$ represents the classical bit 0 (the answer *NO* or the truth-value *Falsity*), while the vector $|1\rangle$ represents the classical bit 1 (the answer *YES* or the truth-value *Truth*). According to the standard interpretation of quantum superpositions, the *quantum perhaps* $|\psi\rangle$ represents an information that might give rise to the answer *NO* (corresponding to the state $|0\rangle$) with probability $|a|^2$ and might give rise to the answer *YES* (corresponding to the state $|1\rangle$) with probability $|b|^2$.

What about systems of n qubits (n-quregisters)? As we have seen, an n-quregister is a possible state of a composite system, consisting of n particles. Following the usual quantum theoretic formalism, the mathematical environment for such a system can be represented as a special product (called *tensor product*) of n two-dimensional Hilbert spaces. On this basis, the mathematical representative of an n-quregister is identified with a unit-vector of the product space

$$\underbrace{\mathbb{C}^2 \otimes \ldots \otimes \mathbb{C}^2}_{n-times}.$$

[4]An *orthonormal basis* of $\mathbb{C}^2$ is a set of two orthogonal vectors of length 1.

Hence, the general form of an n-quregister is usually written as follows:

$$|\psi\rangle = \sum_i c_i |x_{i1}, \ldots x_{in}\rangle,$$

where x_{i_j} is a classical bit, while $|x_{i_1}, \ldots x_{i_n}\rangle$ is an element of the canonical basis of the product space $\underbrace{\mathbb{C}^2 \otimes \ldots \otimes \mathbb{C}^2}_{n-times}$, representing in this framework a classical register (a sequence of n bits).

Quregisters are pure states, hence *maximal pieces of information*, that cannot be consistently extended to a richer knowledge. In quantum computation, one cannot help referring also to non-maximal pieces of information that correspond to *mixtures of quregisters* (briefly, *qumixes*). These are mathematically represented by *density operators* ρ of a space $\underbrace{\mathbb{C}^2 \otimes \ldots \otimes \mathbb{C}^2}_{n-times}$. As expected, quregisters can be described as special cases of qumixes.

2. A holistic semantics suggested by quantum computation

We will now sketch the basic intuitive ideas of quantum computational semantics.[5] The starting point can be described as a natural generalization of classical logic. Let us refer to an "information-theoretic" presentation of classical semantics (and of classical circuit theory). In this framework, sentences are supposed to denote classical bits (either 1 or 0), while the Boolean connectives (*not*, *and*, *or*) represent *logical gates*: functions that permit us to process information. The sentences of *quantum computational logics* (**QCL's**), instead, are supposed to represent quantum pieces of information that are generally uncertain: systems of *qubits* (*quregisters*), or more generally *mixtures* of quregisters (*qumixes*). At the same time, the logical connectives are interpreted as *quantum logical gates*.

What are exactly quantum logical gates? As we have seen, in quantum theory the dynamic evolution of quantum objects is governed by the Schrödinger-equation. Accordingly, for any times t_0 and t_1, a pure state $|\psi(t_0)\rangle$ of an object at time t_0 is transformed into another pure state of the same object at time t_1 by means of a unitary operator $U[t_0, t_1]$ (which represents a reversible transformation):

$$|\psi(t_1)\rangle = U[t_0, t_1](|\psi(t_0)\rangle).$$

Quantum logical gates (briefly, *gates*) are special examples of unitary operators that transform quregisters into quregisters in a reversible way. Hence, from an intuitive point of view the application of a sequence of gates to

[5]For technical details see: M. L. Dalla Chiara, R. Giuntini, R. Leporini, "Logics from quantum computation", *International Journal of Quantum Information* **3**, pp. 293–337, 2005; M. L. Dalla Chiara, R. Giuntini, R. Leporini, "Compositional and holistic quantum computational semantics", *Natural computing* **6**, pp. 113–132, 2007; M. L. Dalla Chiara, R. Giuntini, R. Leporini, G. Toraldo di Francia, "Quantum computational logics and possible applications", *International Journal of Theoretical Physics* **44**, pp. 44–60, 2008.

an input-quregister can be regarded as the dynamic evolution of a quantum object that is processing a given amount of quantum information. By definition gates are unitary operators whose domains consist of vectors of convenient Hilbert spaces. However they can be naturally generalized also to qumixes.

From an intuitive point of view, a quantum information-process can be naturally represented as a kind of *quantum epistemic tree* from an initial state of knowledge $\rho(t_o)$ to a final state of knowledge $\rho(t_1)$:

- $\rho(t_1)$

  $\Uparrow$
- ....

  $\Uparrow$
- $\rho(t_0)$

Although the superficial form of a quantum epistemic tree is that of a *linear process*, the deep structure is essentially *parallel*. For, any qubit $|\psi\rangle$ generally gives rise to a *branching*:

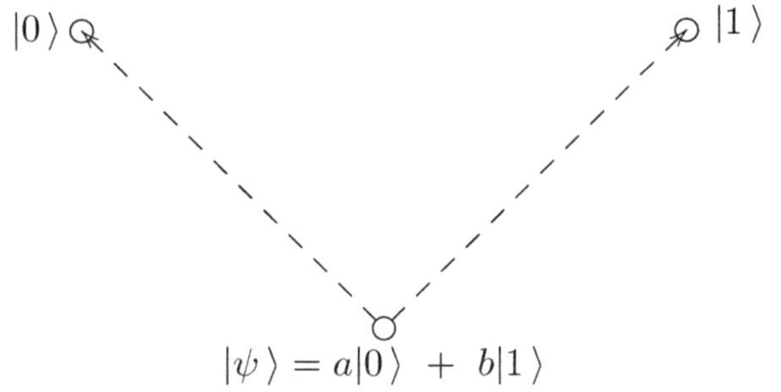

In other words, a state of knowledge represented by a quantum superposition $|\psi\rangle$ reflects, at the same time, two parallel epistemic paths: the first one leads to the answer *YES*, while the second one leads to the answer *NO*. From an intuitive point of view, *quantum computations* can be regarded as characteristic examples of quantum epistemic trees.

As is well known, in any semantic characterization of a given logic, the basic concept is represented by the notion of *model* (or *interpretation*) of the language. For instance, in classical semantics a *model* of a sentential language is a homomorphic map that assigns a classical bit to any sentence, by interpreting the logical connectives as the corresponding Boolean functions.

In the holistic semantics of **QCL's** a *model* (or *interpretation*) of the language is a map `Hol` that assigns to any sentence A a qumix that represents the informational meaning of A:

$$A \mapsto \mathtt{Hol}(A).$$

As expected, any model `Hol` shall preserve the logical form of the sentences, by interpreting any connective $\circ$ of the language as the corresponding gate G°. Furthermore, the qumix $\texttt{Hol}(A)$ should live in a Hilbert space whose dimension depends on the logical form of A. The simplest examples of sentences are *atomic* sentences, which cannot be decomposed into more elementary sentences (say "2 is prime"). Accordingly, the meaning of such sentences shall live in the simplest Hilbert space: the two-dimensional space $\mathbb{C}^2$ (where all qubits are located). A molecular sentence with n occurrences of atomic sentences can be regarded as a linguistic description of a composite physical system consisting of n particles. In fact, we need n particles in order to carry the information that is expressed by our molecular sentence. On this basis, it is natural to assume that the meaning of such a sentence lives in the n-fold tensor product of $\mathbb{C}^2$.

The holistic features of our semantics depend on the fact that any model `Hol` assigns to any sentence A a *global* meaning, that cannot be generally inferred from the meanings assigned by `Hol` to the atomic parts of A. What happens here is just the opposite with respect to the standard behaviour of compositional semantics: $\texttt{Hol}(A)$ determines the meanings of all its parts, which turn out to be essentially *context-dependent*. As a consequence, any sentence may receive different meanings in different contexts. Going from the *whole* to the *parts* is here possible because all logical operations are reversible: one can go back and forth without any dissipation of information!

A fundamental role in this semantic game is played by the notion of *entanglement*, which is mathematically based on the characteristic properties of tensor product. From an intuitive point of view the basic features of an *entangled state* $|\psi\rangle$ can be sketched as follows:

- $|\psi\rangle$ is a maximal information (a pure state) that describes a composite physical system S (say, a two-electron system);
- the information determined by $|\psi\rangle$ about the parts of S is non-maximal. Hence, the states of the whole system is a pure state, while the states of the parts (which are determined by the state of the whole and are usually called *reduced states*) are proper mixtures. It may also happen that the state of the composite system (although representing a maximum of information) describes the parts as essentially indiscernible objects, that cannot satisfy any characteristic individual property. One obtains, in this way, an apparent violation of Leibniz' indiscernibility principle.

Entangled states play an essential role in all *EPR-situations*. For instance, in the case of our metaphorical description of the EPR-paradox, *Sarah* and *Susan* are typically entangled. Before *Oswald*'s measurement, the state of the composite system (*Sarah* + *Susan*) is a pure state that assigns probability $\frac{1}{2}$ to the two following events:

- *Sarah has dark hair and Susan has light hair*;
- *Sarah has light hair and Susan has dark hair*.

A maximal information about the composite system (*Sarah* + *Susan*) determines two reduced states, representing two partial pieces of information about the component parts (*Sarah*, *Susan*). However, such partial pieces of information cannot be represented by pure states. As a consequence, the two individual histories of *Sarah* and of *Susan* turn out to be necessarily ambiguous, before *Oswald*'s measurement.

Entanglement-phenomena can be naturally used to model some typical holistic semantic situations in the framework of our quantum computational semantics. We can consider entangled quregisters that are meanings of molecular sentences. As an example, consider a conjunctive sentence having the form

$$C = A \wedge B.$$

The following situation is possible:

- the *meaning* $\mathtt{Hol}(C)$ of the conjunction C is a quregister, which represents a maximal information (a pure state);
- the meanings of the parts (A, B) are quantum-entangled and cannot be represented by two pure states (two quregisters).

We can say that the *sharp meaning* of the conjunction determines two *ambiguous meanings* for the parts (A, B), which are represented by two mixed states. In other words, *the meaning of the whole determines the meanings of the parts, but not the other way around.* In fact, one cannot go back from the two ambiguous meanings of the parts to the quregister representing the meaning of the whole. The mixed state (i.e. the reduced state) representing the ambiguous meaning of A (of B) can be regarded as a kind of *contextual meaning* of A (of B), determined by the *global context*, which corresponds to the quregister $\mathtt{Hol}(A \wedge B)$ (the meaning of the conjunction $A \wedge B$).

The quantum computational semantics is strongly Hilbert-space dependent. As a consequence, applications to fields (far from the quantum world) where Hilbert spaces do not play any significant role, seem to be somewhat unnatural. However, by abstracting from the Hilbert-space formalism, one can develop an abstract version of quantum holistic semantics that is Hilbert-space free[6]. In this framework, quregisters and qumixes (representing maximal and non-maximal pieces of information, respectively) are dealt with as special kinds of *intensional objects* with growing complexity, which reflects the logical form of possible sentences. Accordingly, an abstract notion of *reduced information* permits one to define *contextual meanings* in an appropriate way (like in the concrete quantum case). This abstract quantum-like semantics seems to represent a flexible tool that might be naturally applied to a number of different fields (including a formal analysis of natural languages and of the languages of art).

---

[6]See M. L. Dalla Chiara, R. Giuntini, A. Ledda, R. Leporini, G. Sergioli, "Entanglement as a semantic resource" *Foundations of Physics* **40**, pp. 1494–1518, 2009.

As an example, we can try and apply such a semantic analysis to the poem *L' Infinito*. Let us (artificially) decompose the poem into the two following sentences:

- B = *'l naufragar m'è dolce in questo mare (drowning in this sea is sweet to me)*
- A = the poem *L' Infinito* without the last verse.

Thus, we obtain:

$$L'Infinito = A \wedge B.$$

Our semantics describes how the contextual meaning of the last verse B (an ambiguous meaning) is determined by the global meaning of the whole poem.

CHAPTER 4

Computing and thinking: does the human brain behave as a quantum object?

1. Computing by a Turing machine

Quantum computation has stimulated a general debate about the critical relationships between two kinds of human activities that we normally call "thinking" and "computing". In particular, the three following questions have been object of discussion:

1. Do quantum computers only represent machines that are more efficient and more speedy than classical computers? Or will we instead find examples of computation, impossible for a classical computer, that a quantum computer could execute, in a finite time? In other words, do quantum computers force us to radically transform our intuitive ideas about the very notion of *computer*?
2. To what extent can quantum computers simulate human *reasoning* and *thinking* more faithfully than classical computers?
3. Is it sensible to guess that the human brain behaves like a kind of quantum object?

In order to discuss these questions it will be expedient to recall first the basic properties of the notion of *Turing machine*, which represents an abstract model for the intuitive concept of computer (proposed by the great logician and mathematician Alan Turing, in 1935). The starting point of Turing's theory can be described as an attempt to give a universal answer to the question: *what does exactly computing mean*? The idea is that the agent who performs a computation might be represented either by a human person or by a machine. In both cases, the characteristic features and activities that play a fundamental role in any computation-procedure can be sketched as follows:

a) *reading and writing*;
b) *going through different memory-states*;
c) *acting during a finite and discrete time.*

On this basis, the concept of *Turing machine* can be defined as a system **M** that satisfies the following properties:

1. **M** is equipped with a (potentially infinite) *tape*, divided into *discrete cells*, where at most one symbol can be printed.
2. **M** makes use of an *alphabet* consisting of a finite number of symbols.

FIGURE 1. A Turing machine

3. **M** has the capacity of *moving* along the tape, of *seeing* one cell at a time, and of performing one of the following operations: *going to the right*, *going to the left*, *standing still*, *printing* or *erasing* a single symbol (where *erasing* can be identified with *printing the empty symbol*).
4. **M** assumes different *memory-states* that are functions of the operations performed step by step. Conventionally, such states can be identified with some abstract objects, representing pieces of information that are possibly "known" by our machine.

This is the structure of any Turing machine (Fig. 1).

At the same time, each example of a particular machine will be determined by an instruction-system that represents the *program* of the machine in question. Any elementary instruction will have the following form:

> if you *are* in a given *memory-state* and you *see* a given *symbol*, then you shall *move along a given direction* (either moving to the right or moving to the left or standing still), soon after you shall *print* a given symbol, finally you shall *assume* a given *memory-state*.

Such an instruction can be schematized by a sequence of five abstract objects:

$$(St_i,\ Symb_s,\ D,\ Symb_p,\ St_f),$$

where St_i, $Symb_s$, D, $Symb_p$, St_f represent respectively: the initial memory-state, the symbol that is seen, the movement's direction, the symbol that is to be printed, the final memory-state.

Accordingly, an instruction-system can be represented as a sequence of quintuples, having the form described above. As expected, any program shall be *consistent* and *uniquely determined*. In other words, the following condition must be satisfied: whenever the machine is in a given state and sees a given symbol, the program shall give a unique order: *perform these operations and finally assume the following final state!* Apparently, one is dealing with a perfect deterministic situation.

As happens in the case of physical systems, Turing machines also evolve in time, assuming a different *configuration* at each instant. From an intuitive point of view, the machine-configuration at time t represents the *general state* of our machine at that particular time. As a consequence, it is sensible to require that any configuration is determined by three components:

1. the *position* of the machine at time t, which corresponds to the cell that our machine sees at that particular time.
2. The *condition* of the machine at time t, which is determined by all that has been printed on the tape until time t.
3. The memory-state of the machine at time t.

At this point, we have at our disposal all that is needed in order to answer the question: "what does computing mean?" As a first approximation, we can say: *for a machine* **M** *computing means passing from one configuration to another one, according to the instructions of its program, during a finite time*. For instance, what does it mean that **M** is able to calculate numerical sums (as our personal computers do, provided that the numbers to be summed are not "too large")? In the case of sums, the natural answer will be the following:

> for any pair (m, n) of natural numbers there exists a computation of the machine **M** whose initial configuration contains the arguments m and n, while the final configuration contains the value $m + n$.

In other words, whenever **M** receives a given input, then **M** is able to determine the output, after a finite number of steps.

On this basis, one can propose the following general definition of *Turing-computability*:

> a numerical function f is called *Turing-computable* if and only if there exists a Turing machine **M** that calculates the value of f for any possible choice of arguments for f.

Since, by definition, computations are always finite, any machine that executes a given computation will finally halt. Metaphorically, we can imagine that our machine says: "well, I have finished my job!" On this basis, one can assert that any computation corresponds to a process where a given Turing machine finally halts. Technically, a halting-configuration corresponds to an instruction whose final memory-state is a special state conventionally called the *halting-state*.

How many are all possible Turing machines? Of course, infinitely many! In fact, any instruction-system corresponds to a particular machine. In spite of this, interestingly enough, one can prove the existence of a *universal Turing machine* that is able to simulate all that any particular Turing machine can do. This is a consequence of the fact that our general definition of Turing machine does not assume any limitation concerning time, space and memory of our machines. The only request is that any computation-process be finite.

Apparently, Turing machines represent pure mathematical objects that are fully independent of any technology. Turing's theory is not concerned with hardware. At the same time all concrete computers that have been constructed so far (from our modest note-book to the powerful *Deep Blue*) are examples of Turing machines. An important question arises at this point: to what extent is the general notion of Turing machine strong enough in order to faithfully describe all possible (present and future) computers? Are we entitled to identify the intuitive notion of *computation* with the formal notion of *Turing-computation*? This question is positively answered by the so called *Church-Turing thesis*, which has been independently proposed by the logician Alonzo Church and by Turing himself.

Apparently, the Church-Turing thesis cannot be proved, since it does not have the logical *status* of a possible theorem. In fact, what the thesis asserts is a typical adequacy-statement, that compares an intuitive non-rigorous notion (*computability*) with a precise formal concept (*Turing-computability*). While it cannot be proved, the thesis in question might be refuted. To this aim, it would be sufficient to construct a computer that is able to perform some computations, which are, in principle, impossible for a Turing machine. But are there any mathematical operations that are, in principle, impossible for a Turing machine? Does classical mathematics show the existence of numerical functions that are not computable? Or, shall we instead pretend that any mathematical problem is solvable by means of a convenient computation, provided that the time at our disposal is sufficiently long?

Let us refer to the set of all natural numbers (0,1,2,...) and let us consider the operations that can be defined on this set. Among them we will have, of course, the well-known elementary operations (like sum, product, exponentiation, and so on). There is a deep relationship that connects *functions* with *properties* and *relations*. For the sake of simplicity, let us only consider the case of numerical properties (like *even*, *odd*, *prime*, and so on). One can easily see that each property P uniquely determines a function f_P (called the *characteristic function* of P). From an intuitive point of view, f_P can be regarded as the function that gives only two possible answers: 1 (*YES*) or 0 (*NO*). The answer will be: *YES*, when function f_P

is applied to a number that satisfies property P; *NO*, otherwise. For instance, the characteristic function of the property *even* will assign value 0 to the number 0, value 0 to the number 1, value 1 to the number 2, and so on. Suppose now that the characteristic function f_P of a property P is Turing-computable. This means that there exists a machine **M** that is able to answer either *YES* or *NO* to any question asking whether a particular number n satisfies property P (after a finite number of steps). In such a case we can say that property P is *decidable*. In other words, for any choice of a particular number n, one can decide in a finite number of steps whether our number satisfies P. Viceversa, the decidability of P automatically determines the computability of the characteristic function f_P. As a consequence, we can conclude that *computability* and *decidability* represent two deeply correlated concepts. Should we discover the existence of a non-decidable numerical property P, we would immediately obtain that there exists at least one function (the characteristic function f_P) that is non-computable. And viceversa.

2. A "masquerade ball" in the proof of Gödel's theorems

Another important relationship connects the concept of *decidability* with the concept of *provability* in a particular formalized mathematical theory.[1] The theory that plays such "privileged" role is represented by a formalized version of arithmetic, according to the axiomatization proposed by Giuseppe Peano at the end of the 19th Century. By simplicity, we will briefly call this theory (which we will suppose formalized in the framework of elementary classical logic) *PEANO*. One of the special "virtues" of *PEANO* is the following: it faithfully describes *all and only* the numerical properties that are decidable. For, the following relation holds:

> any numerical property P is decidable if and only if P is syntactically decided in *PEANO*. In other words, whenever a number n satisfies P, then *PEANO* proves a sentence asserting that "n satisfies P". Viceversa, whenever n does not satisfy P, then *PEANO* proves the negation of the sentence in question.

As a consequence, one immediately obtains that any form of undecidability or non-computability would give rise to a syntactical limitation of the theory *PEANO*: there should be at least one arithmetical sentence, expressed in the language of *PEANO*, that would escape the control of *PEANO*'s proof-theoretic "powers". Hence, the question arises: are there any arithmetical sentences that *PEANO* is not able to control syntactically? An answer to this question has been given by the celebrated *incompleteness theorems* proved by Kurt Gödel in 1931.

---

[1]We recall that formalizing a theory means determining in a precise way its syntactical components. In particular, one shall rigourously define the following notions: *well-formed expression*, *sentence*, *axiom*, *proof*, *theorem*.

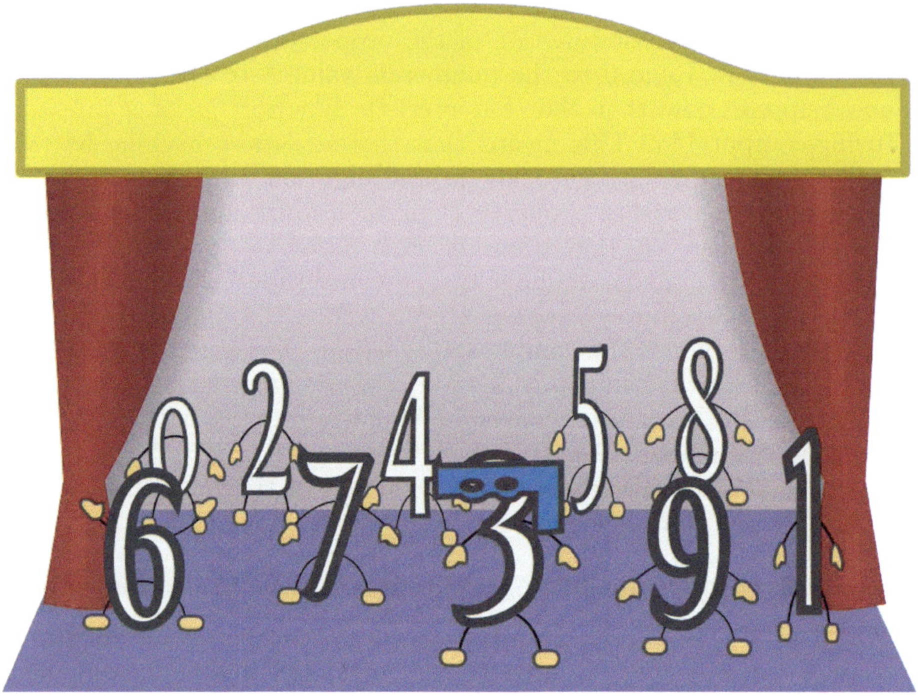

FIGURE 2. A masquerade ball on a numerical stage

In order to illustrate the intuitive content and the epistemological consequences of Gödel's theorems, it may be expedient to use a "theatrical metaphor". Let us imagine an ideal stage, where all natural numbers (0,1,2,...) are located, while some of them are wearing a kind of "syntactical mask". Such a "masquerade ball" can be performed in the following way: by using a special technique based on an *arithmetization of the language*, all alphabet-symbols, all expressions and all finite sequences of expressions of the arithmetical language are one-to one associated to particular numbers. In this way, some numbers acquire a "double role": their "genuine" numerical role and the role that is determined by their syntactical mask. For instance, we might choose a codification according to which number 3 wears the mask of the logical connective *negation* ($\neg$). (Fig. 2).

On this basis, all syntactical properties and relations are transformed into numerical properties and relations. For instance, the property "to be a theorem of *PEANO*" becomes a numerical property, quite similar to ordinary numerical properties like "to be an even number" or "to be a prime number". This gives rise to some "dangerous" self-reference phenomena that generally cause a sense of psychological uneasiness, a kind of "cognitive vertigo", as happens when we look at the so called *impossible pictures*.

FIGURE 3. Gödel's sentence that asserts its own unprovability

An important example of a self-referring sentence (expressed in the language of *PEANO*), which after Gödel's proof has been termed *Gödel's sentence* (often indicated by the letter G) asserts:

I am not a theorem.

As a consequence, by arithmetization of the language, there must be a number g that appears on the arithmetical stage with a quite intriguing "mask", where is written "I am not a theorem" (Fig. 3). One is dealing with a "numerical character" that has many properties in common with the celebrated *Liar*, who asserts "I am telling the falsity", giving rise to the *Liar's paradox*.

Under the hypothesis that *PEANO* is a consistent theory, one can prove that Gödel's sentence G cannot be either proved or disproved in the theory, just because it asserts its own unprovability. On this basis, Gödel's *first incompleteness theorem* can be formulated as follows:

> *any consistent scientific theory, that includes arithmetic (and satisfies some canonical syntactical conditions) is* logically incomplete, *since it expresses some sentences that are syntactically undecidable (neither provable nor disprovable) in the theory.*

Now, we know that according to classical semantics, all sentences are either true or false (*tertium non datur!*). Hence a question naturally arises: is the sentence G true or false? Having proved that G is unprovable, we shall conclude that G is true. Indeed, G that asserts its own unprovability is really unprovable! Thus, there exist examples of mathematical truths that cannot be proved. Against a traditional opinion (often defended in science and in philosophy) the concept of *mathematical truth* cannot be reduced to the syntactical concept of *provability*.

Interestingly enough, the sentence G turns out to have another important intuitive content: just because G asserts it own unprovability, G asserts that "some sentences of the theory are unprovable"; hence, "the theory is consistent". In fact, should the theory be inconsistent, then any sentence would be provable, because contradictions logically imply all sentences expressed in the language (*ex absurdo sequitur quodlibet!*). As a consequence, we can say that our number g is entitled to wear on the numerical stage also another mask, where is written "the theory is consistent". On this basis, Gödel's *second incompleteness theorem* can be obtained as a corollary of Gödel's firs incompleteness theorem:

> *any consistent scientific theory, that includes arithmetic (and satisfies some canonical syntactical conditions) can express but cannot prove its own consistency.*

As we have seen, Gödel's sentence G is at the same time undecidable and true. This gives rise to the following situation:

- the human mind is able to "see" that G is unprovable;
- the universal Turing machine does not halt when we ask the question "is G provable?".

Prima facie, such situation seems to lead to the conclusion that the human mind is able to grasp some mathematical truths that are inaccessible for the universal Turing machine, and also for any form of artificial intelligence (assuming the validity of the Church-Turing thesis).

However, things are a little bit more complicated. Although each particular consistent theory (that includes arithmetic and satisfies some canonical syntactical conditions) necessarily expresses some sentences that are undecidable in the theory itself, one can prove the existence of an infinite sequence of theories such that any arithmetical problem finds a solution in a given element of the sequence. For instance, the sentence G itself (which is improvable in $PEANO$) can be proved in a stronger theory (like an axiomatic version of *set theory*). Apparently, what we call "mathematical intuition" cannot be assimilated to a unique formalized theory. Human minds seem to be simultaneously weaker and stronger than any particular theory or Turing machine. They are weaker because their computation-capacities cannot be compared with the capacities of the universal Turing machine. Obviously, no mathematician could prove *all* theorems of $PEANO$. At the same time, human minds are also stronger than any particular theory or machine. For, humans are equipped with a peculiar ability that permits them to "jump" through different theoretic contexts, going back and forth and considering different theoretic *levels* (the *object-theory*, the *metatheory*, the *meta-metatherory*, and so on). And such cognitive processes are realized by acting, at the same time, on different parallel paths. Just this ability is responsible for our capacity of "seeing" self-reference situations, which may instead create logical conflicts whenever we are referring to a unique theory.

3. The quantum and the brain

Let us now go back to our initial questions concerning some epistemological consequences of the theory of quantum computation. Apparently, our first question leads us to discuss the validity of the Church-Turing thesis: are quantum computers only more efficient than classical computers? Or, are they, instead, essentially stronger? In other words, is there any problem that a quantum computer could solve, while the same problem is, in principle, unsolvable for the universal Turing machine, quite independently of the time at our disposal? This question, which has raised many interesting discussions, seems to be still open (although one can sometimes find in the literature "opinions" that are tentatively presented as "theorems").

The crucial difference between the ideal computers that can be called *quantum Turing machines* and classical Turing machine is very clear: a quantum Turing machine can be described as a quantum superposition of many classical Turing machines. Now, it is well known that in classical contexts, any parallel computation can be simulated by a linear sequence of Turing machines, working one after another. Consequently, even the behaviour of classical parallel computers can be assimilated to the behaviour of the universal Turing machine. Is such a simulation also possible in the case of computational situations, where parallelism is essentially founded on quantum superpositions? As we have learnt, quantum superpositions give rise to the characteristic interference-phenomena. As happens in the case of the two-slit experiment, where an electron goes at the same time through slit 1 and slit 2 (interfering with itself), in a similar way, during a quantum computation different computational paths are followed at the same time, giving rise to interference phenomena. Is it possible to *linearize* a process of this kind? So far, no satisfactory answer to this critical question is known.

Is it interesting to compare quantum Turing machines with the characteristic parallel behaviour of the human brain? Can we reasonably guess that quantum computers are somehow able to simulate the computational procedures and perhaps even the "thought-capacities" of human minds?

A suggestive hypothesis, which raised a number of interesting discussions, has been proposed by the physicist Roger Penrose.[2] According to

---

[2]See R. Penrose, *The Emperor's New Mind*, Oxford University Press, Oxford, 1989, and R. Penrose, *Shadows of the Mind*, Oxford University Press, Oxford, 1994.

Penrose, the non-computational behaviour of the human brain, which is able to grasp some non-computable mathematical truths (like Gödel's sentence G) might depend on some quantum phenomena. The main idea is to investigate the neuron-inner structure, in particular the *microtubules* (inside the axons), which are responsible for the transmission of nerve signals. Being completely isolated from the external environment during a long time and because of their size, microtubules might behave like *Bose-Einstein condensates* (i.e. systems consisting of many particles, all in the same state).

But why should a quantum behaviour explain the non-computational aspects of human brain-activities? According to Penrose, a special role is played by the *collapse of the wave-function.* As we have learnt, such principle determines casual, unpredictable and non-computable transformations in the set of all possible states of a quantum object. Nevertheless, Penrose believes that the present status of quantum theory is not sufficient to explain how quantum phenomena intervene in the activity of the human brain. What is needed is a new physical theory where the Schrödinger-equation and the collapse of the wave-function can be reconciled in a satisfactory way.

Penrose's proposal has raised a number of criticisms. Some radical objections have been put forth by the biologists' community, according to which microtubules are a kind of objects that are too big in order to be sensitive to quantum effects. We can also observe that, strangely enough, a contradiction seems to emerge from Penrose's overall ideas that defend at the same time the following claims:

1. any quantum physical system (whose number of degrees of freedom is finite) can, in principle, be simulated by a quantum computer;
2. quantum computers are examples of Turing machines, because all forms of parallel computation can, in principle, be linearized;
3. there is something in the human brain that is deeply non-computational: *thinking* and *computing* are different activities;
4. a special kind of quantum phenomena could be responsible for the "thought-capacities" of the human brain.

Apparently, these four claims cannot be asserted at the same time, because they give rise to a possible logical conflict, although each of them appears reasonable from a particular point of view. In fact, suppose that some quantum phenomena are responsible for the "thought-capacities" of the human brain (4.). Then, it seems reasonable to expect that the human brain can be described as a special example of a quantum object. However, any quantum system (whose number of degrees of freedom is finite) can, in principle, be simulated by a quantum computer (1.). At the same time, quantum computers are examples of Turing machines (2.). As a consequence, the "thought-capacities" could be simulated by a Turing machine. Hence, it is not true that there is something in the human brain that is deeply non-computational (against 3.).

Interestingly enough, the relationships between the quantum world and the activity of the human brain can be also investigated, by assuming, as a starting point, a different attitude. One is dealing with a perspective that is not aimed at isolating in the brain sufficiently small physical objects, which might determine quantum effects. The attention is, instead, shifted to some universal features of the logical and mathematical formalism of quantum theory, that can be successfully exported even to fields that are deeply far from the micro-world. This is the perspective that we are going to follow in the next Chapters, where we will try and apply some semantic ideas, originated in a quantum-framework, to a formal analysis of music.

CHAPTER 5

The musical monad and its context

The sound-world is a typically *relational* world, whose behaviour is quite different from colours. Generally, we cannot associate precise meanings to single notes or sounds. In a sense, all single notes appear to us semantically equivalent. Is it sensible to assert that musical intervals (rather than single notes) represent a kind of *musical monads*?

As is well known, in the framework of musical theory an *interval* is defined as the *distance between two pitches*. On this basis, it is customary to distinguish *melodic* from *harmonic* intervals. While harmonic intervals (which represent the fundamental nucleus of harmony) are characterized by the simultaneous presence of two different pitches, a melodic interval is based on a sequence thereof. However, one cannot assert that any sequence of two sounds represents a melodic interval. In order to perceive a given sequence as a melodic interval, the two sounds should be performed during a "sufficiently short" time: it is essential that the memory of the first sound is still preserved when the second sound is going to be performed. As Roberto Lupi used to say, any melodic interval lives between a note that is dying and an another note that is coming to life.

In this perspective melodic intervals can be regarded as a subjective response of human perception to some sound-phenomena. This happens when a listener (or a performer) realizes a relationship between two sounds (with different pitches) that are performed during a sufficiently short time. One is dealing with a characteristic situation that concerns all musical experiences. In this sense one can say that melodic intervals represent a kind of *atomic monads* of the universe of music.

A synthetic representation of the basic intervals (of the tempered scale-system) is illustrated in Fig.1 and Fig.2 (where the notations "2^+" and "2^-" mean "major-second interval" and "minor-second interval", respectively; in a similar way, in the case of the other numbers. The notations "4^{aug}" and "4^{dm}" mean "augmented fourth-interval" and "diminished fourth-interval", respectively).

Each interval of the tempered scale, from the minimal interval represented by the *minor second* (say, $C - D$ flat) to the *octave* (say, $C - C^8$), has an intrinsic expressive quality, which is naturally perceived in the framework of our Western tradition. In order to illustrate this idea, we will now refer to some significant examples that belong to the lyric, symphonic and chamber music repertoire.

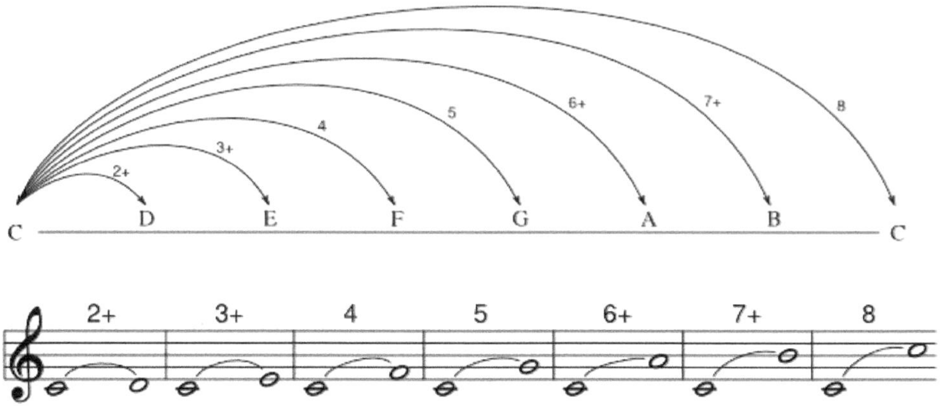

FIGURE 1. The basic melodic intervals

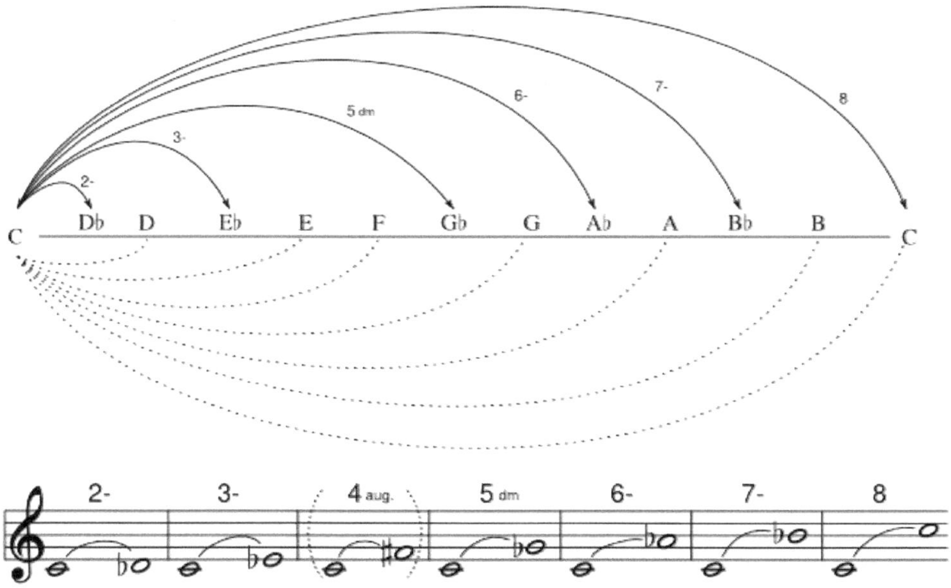

FIGURE 2. Some other intervals

In the famous "Esultate!" (Rejoice!) of Verdi's *Otello* the *major-sixth interval* fully expresses the joy and the enthusiasm determined by Otello's return, at the very beginning of the opera. The expressive power is here strongly emphasized by the semantic correspondence between music and text:

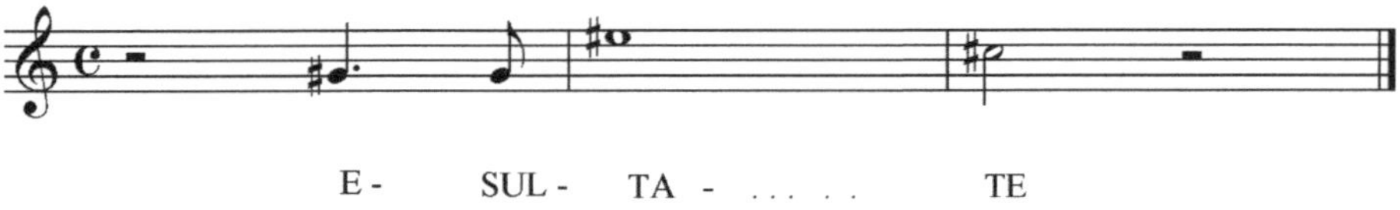

Also the euphoric toast in *La Traviata*'s Act I is expressed by means of repeated major-sixth intervals:

An example that represents an opposite semantic situation is the *incipit* of Monteverdi's *Arianna's Lament.* Here two *minor-second intervals* (an ascending and a descending one) express the dramatic content of the text ("Let me die"):

Third-intervals, in turn, represent a characteristic aspect of most lullabies, *canciones de cuna* and Christmas songs. In fact, such intervals seem to be often connected with serene atmospheres or consoling psychological attitudes. A significant example is the *incipit* of the last movement in Beethoven's 6th Symphony ("Pastorale"):

Another celebrated example is the theme of the second movement in Dvořák's 9th Symphony "From the New World":

The *major-seventh* is the interval that probably expresses a maximal tension-degree, due to the position of the two sounds in the scale (for instance, $C - B$). As an example, we can mention the finale of Act I in Puccini's *La Fanciulla del West* (*The Girl of the Golden West*). Here, a sequence of descending major-seventh intervals seems to suggest Minnie's surprise and emotion, after Johnson's flattering in the previous duet:

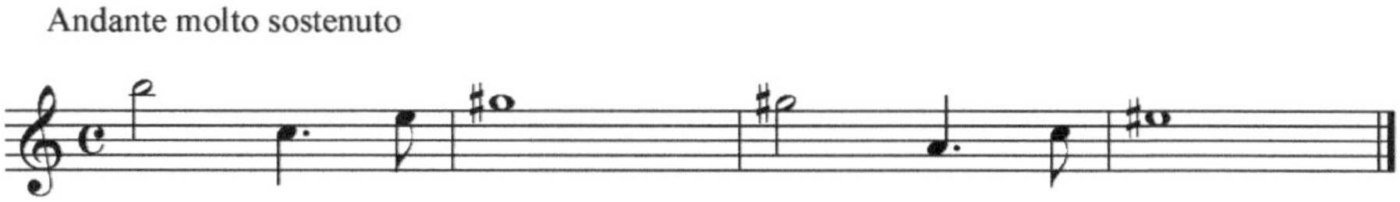

Another significant example where the repetition of ascending major-seventh intervals plays an important role is the theme of the duet "O Terra addio" ("Farewell, earth") in the last Act of Verdi's *Aida*:

The attractor-function of the *tonic* (the first degree of the scale) versus the *leading note* (the seventh degree) is here straightforward.

Some longing seventh-intervals give a characteristic expressive colour to the third movement of Schumann's Piano Quartet *Op.47*, where the melodic design of the main theme is based on several seventh-intervals, which are frequently repeated on different degrees of the scale:

The *octave-interval* expresses a kind of apodictic assertion, due to the fact that the first note (say, C) is repeated at the next octave (say, C^8). Accordingly, such interval seems to "close" all possible intervals into a kind of circle.[1] Two significant examples from Mozart are the *incipit* of the

[1]It is worthwhile recalling that the frequency of the higher sound of an octave-interval is the double frequency of the lower sound. Such interval has a very important role in the harmonic sound-sequence.

"Haffner" Symphony *K. 385*:

and the first measures of the Quartet in *D* minor *K. 421*:

An interval that suggests a kind of "opening" towards the external world is the *ascending fifth* (say, $C - G$). This happens whenever we move from the *tonic* to the *dominant*. The term "dominant" itself (which denotes the fifth degree of the scale) seems to evoke a kind of "suspended domain" in the interval-universe. Generally, an ascending fifth opens, announces or prepares some new sound-events. As an example, we might mention the Introduction of Haydn's "London" Symphony *No. 104*:

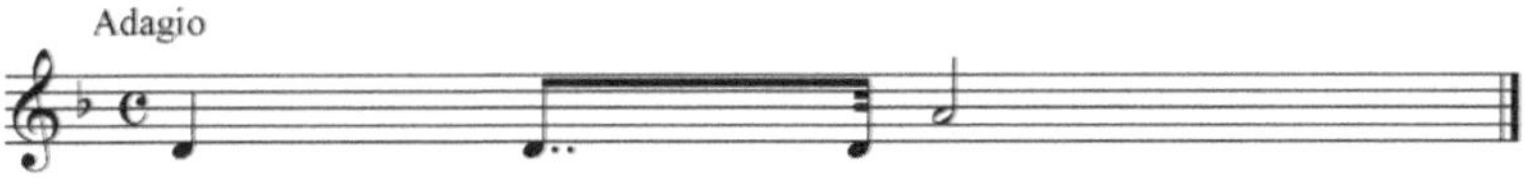

The *descending fifth* (say, $G - C$) has, instead, an opposite character: such interval seems to move from outside to inside, somehow closing a kind of ideal speech. We need only think how often the conclusion of a composition is based on a descending fifth, that clearly stresses the closing.

A less characterized interval is the *fourth* (both in the ascending and in the descending version). As well as the fifth, the fourth also suggests an opening (in the ascending case) and a closing (in the descending case); however the dynamic intensity is less powerful. As an example, we might mention the delicate *incipit* of many phrases in Schumann's *Träumerei*, from the *Kinderszenen Op. 15*:

FIGURE 3. A minor-second interval for Arianna and for Figaro

or the recurrent emerging of the main theme in the first movement of Schubert's "Unfinished" Symphony *No. 7*:

The expressive features that characterize the different musical monads (represented by the melodic intervals) may undergo some significant changes in particular contexts. Such semantic transformations may depend on different variables that generally concern the rhythm or the timbre or the speed. For instance, the sad character that we have generally attributed to the minor-second interval is completely transformed in the main theme of the *Ouverture* of Rossini's *Barbiere di Siviglia* (Fig. 3). Here the interval acquires an irresistibly humorous colour, due to a fast *tempo*:

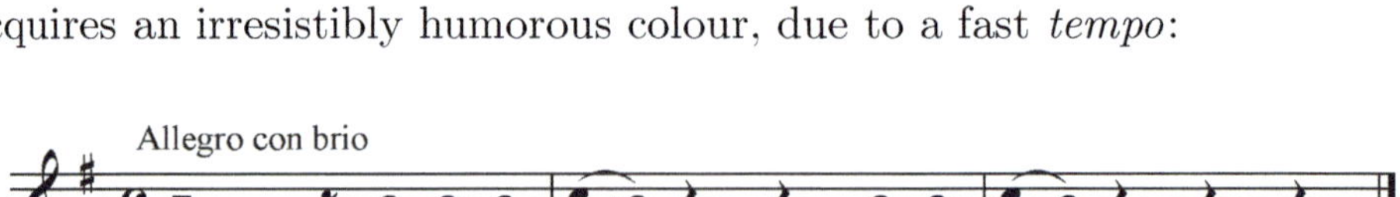

Another example based on the iteration of minor-second intervals is the main theme in the first movement of Mozart's Symphony in *G* minor (*No.*

40 K. 550). Here the *tempo*-choice (*Allegro molto*) mitigates the sad connotation we have recognized in the case of *Arianna's Lament*:

An interesting case is the *incipit* of the Witch Baba-Yaga's episode in Mussorgski's *Pictures at an Exhibition*, where the rhythmic and dynamic violence of descending major-seventh intervals amplifies the characteristic tension of such intervals, giving rise to a special sound-brutality:

Generally, changing the harmonic context where a given interval occurs may determine a different sound-effect. For instance, the minor- third interval $C - E$ flat in the framework of a C minor-harmonic basis, may change its colour and its expressive power, if the lowest note C is replaced by an A flat. Such a harmonic change seems to clear up the two notes, bringing a kind of new light[2]:

The examples we have presented show how each interval has a characteristic semantic connotation, which may also depend on the (musical or extra-musical) context where our interval appears. In the next Chapters we will discuss the general question concerning the *contextuality* of musical meanings, from an abstract point of view.

[2]In this example, one clearly moves from a minor to a major tonality. Notice however that the harmonic change is not the only reason that determines the semantic transformation. Also the position-change of the two notes in the scale has a bearing. In fact, in the first measure of our example the E flat represents the third degree of the scale, while the C is the first degree. In the second measure, instead, the E flat is the fifth degree, while the C becomes the third degree.

CHAPTER 6

Musical scores: a two-dimensional syntax

Musical scores are very complicated examples of symbolic languages. It is interesting to analyze how information is coded by musical scores in comparison with the standard formal languages that are used for scientific theories. The most important differences seem to be the following:

- Formal scientific languages are basically *linear*: *words* and *well-formed expressions* are represented as *strings* consisting of symbols that belong to a well determined alphabet. The relevant syntactical relations can be adequately represented by a *Turing machine*, which refers to a one-dimensional tape.
- Scores, instead, are two-dimensional syntactical objects, which have at the same time a horizontal and a vertical component. Any attempt to linearize a score would lead to totally counter-intuitive results.

From a semantic point of view, the characteristic two-dimensionality of musical notation seems to be significantly connected with the deep parallel structures that have an essential role in our perception and intellectual elaboration of musical experiences. We might recall, in this connection, a happy metaphor proposed by Antonio Damasio, according to which the human brain seems to work like an orchestra![1]

As is well known, music and spoken talks are usually perceived according to different modalities. While a simultaneous superposition of talks generally determines a psychological sense of uneasiness, music instead gives rise to the mysterious "polyphonic pleasure". As an example, let us think of many duets (or terzettos or quartettos) of lyric operas. Generally, a listener perceives the global polyphonic result; at the same time he (she) is able to grasp the different melodic lines as well as the different "thoughts" of all characters who are singing. We might refer to a number of celebrated examples. A significant case is represented by a fragment of a famous duet in the opera *La Traviata*, where Alfredo's father (Germont) tries to convince Violetta that she should leave Alfredo. At the very beginning, Violetta proposes a kind of compromise: "Ah comprendo, dovrò per alcun tempo da Alfredo allontanarmi" ("Ah yes, I understand. If I felt Alfredo for a time..."). This is what she says, by means of a kind of *recitativo*. However,

[1] A.R. Damasio, *Feeling of What Happens: Body and Emotions in Making of Consciousness*, Harcourt, 1999.

Violetta has understood very well that what Germont is asking for is much more serious: the separation shall be very for ever. From the musical point of view, her thoughts and her anguish are not realized by the melodic line of her singing but rather by the dramatic phrases performed by the strings in the orchestra. And the "contradiction" between what Violetta is saying and what Violetta is thinking is significantly expressed by some dissonant chords (for instance, Violetta "says" an A flat and "thinks" an A).

Is it possible (and interesting) to represent a musical score as a peculiar example of a formal language? In a sense, are scores formalizable? One can positively answer to this question, by introducing the notion of formal representation of a musical score. From an intuitive point of view, we can imagine the formal structure of a score as a configuration of symbols that are written on a piece of graph-paper: each row of the paper corresponds to what shall be performed by a particular instrument (say, the first violin); while columns correspond to what shall be played at the same time. Each cell of our paper can be regarded as a container for an atomic information. From the mathematical point of view, such two-dimensional syntactical configurations can be naturally represented by means of some special matrix-like structures. Accordingly, any score-measure can be represented as follows:

$$\begin{pmatrix} Ins_1 : A_{11}^{[r]} \ldots A_{1n}^{[r]} \\ Ins_2 : A_{21}^{[r]} \ldots A_{2n}^{[r]} \\ \ldots \\ Ins_m : A_{m1}^{[r]} \ldots A_{mn}^{[r]} \end{pmatrix}$$

where each row corresponds to a particular instrument (or voice), while columns correspond to what shall be played simultaneously. Any $A_{ij}^{[r]}$ represents a *score-atom*: a piece of information for the i-th instrument at the j-th position of the r-th measure. For the sake of simplicity, one can assume that the number of rows remains constant through the score. Whenever a given instrument *tacet*, we suppose that some appropriate pause-symbols are written for the instrument in question.

The traditional musical notation represents a rather complicated code: in fact, reading a score turns out to be quite difficult for non-professional musicians. The language of music is, without any doubt, much more complex and rich than the standard formal languages used by scientific theories. Roughly, we can identify at least the following classes of symbolic expressions that play a fundamental role:

a) names for the different notes that represent different sound-pitches, (for instance, the diapason-A, corresponding to the approximate frequency of 440 Hertz);
b) names for the different pauses;
c) meter-indications (like $\frac{4}{4}$, $\frac{6}{8}$, and so on);

d) metronomic indications;
e) tempo-indications (like *Allegro*, *Adagio*, *Vivace*, and so on);
f) dynamic-indications (like *piano*, *forte*, *crescendo*, *diminuendo*, and so on);
g) prescriptions about the sound-emission (for instance, *legato*, *staccato*, *pizzicato*, and so on);
h) names for the different instruments and for the different kinds of voice (violin, viola, soprano and so on).

Traditionally, note-names are indicated by referring to the staff-notation and to some conventions that make essential use of the clefs, the accidentals and so on. In this framework, each occurrence of a note-name is associated with a certain rhythmic value (say, $\frac{1}{4}$). A number of indications are given either at the beginning of a movement (for instance, *Allegro*) or at the beginning of a measure or at the beginning of a line (which consists of a sequence of measures). Of course, all initial indications are supposed to be distributed over what is following, if they are not changed in the course of the score. Any formal version of a score shall reflect all these pieces of information, possibly adopting some different conventions. For instance, one could use some arithmetization-techniques, by assuming that note-names are conventionally represented by particular natural numbers. Another useful "trick" is based on the following formal convention: each score-cell is supposed to contain a single note-name (or a single pause-name). Accordingly, all chord-notations turn out to be distributed over different score-rows. To this aim, one shall artificially increase the number of *abstract instruments* (which corresponds to the row-number in the score). Suppose, for instance, that the chords played by the first violin consist of four notes, at most. We shall transform our score into a new (musically equivalent) score, where four first violins are taken into consideration. And we shall repeat such transformation for all score-instruments. Needles to say, one is dealing with a formal transcription that would not help the reading of real performers! But formalization has quite different aims.

A concept that plays an important role for musical languages (as well as for natural and scientific languages) is that of *phrase*. What is exactly a *musical phrase*? Since the 18th Century a rich analytical literature has tried to propose some rules in order to define the critical notion of *musical phrase*. One is dealing, however, of rules that admit a lot of variants; as a consequence it is impossible to refer to a unique precise definition that can be applied to the whole repertoire of tonal music.[2] In fact the notion of musical phrase has been split into different sub-notions: *semi-phrase*, *section*, *member*, *cut*,.... As a result, we are concerned with a variable concept that acquires different meanings in different situations, also in the framework of one and the same historical period. In a sense, all variants and

[2]See, for instance, the repertoire quoted in I. Bent, W. Drabkin, *Analysis*, Macmillan Publisher LTD, London, 1980.

exceptions concerning the size of musical phrases represent essential features of the characteristic compositional ideas of each musician.[3]

How to deal with musical phrases in a formal setting? From a syntactical point of view, musical phrases are similar to *well-formed expressions* of scientific formal languages (symbolic constructions that may acquire a meaning, once given an *interpretation* of the language). There is however a deep difference between musical and scientific languages. The set of all well-formed expressions of a formal language is generally *decidable*: given a sequence of symbols, one can always decide in a finite number of steps whether the sequence in question represents a well-formed expression of the language. As we have just recalled, the concept of musical phrase is, instead, somewhat fuzzy, because the formation-rules are not rigid, while a great number of exceptions is admitted. Furthermore, in many cases, the choice of the significant phrases may depend on the intentions of the performer who proposes a given interpretation.[4]

From an abstract point of view, any formal definition of *musical phrase* shall take into account the following (minimal) characteristic properties of concrete musical phrases (briefly, *phrases*):

- a phrase generally consists of a (small) number of measures.
- A phrase does not necessarily begin at the beginning of a measure and does not necessarily end at the end of a measure.
- Phrases are generally *transversal* with respect to the instruments of the score: an instrument may begin a given phrase, while other instruments will continue it.
- A phrase does not generally concern *all* the score-atoms contained in the measures where parts of the phrase in question appear. Hence, phrases can be formally represented as pieces of score with "holes", which will be also called *empty score-atoms* (and indicated by the symbol ♣). Of course, empty score-atoms (which represent "holes") should not be confused with pause-symbols.

On this basis, musical phrases can be described as special *linguistic objects* that represent fragments of a formal score. Since isolating a phrase in a score means extracting a fragment, we can imagine that the score-cells that are not involved in the fragment in question are empty. Accordingly, a phrase of a formal score can be identified with a sequence of score-columns, where some symbols have been substituted by the hole-symbol ♣. We will require the following natural condition: all the columns that constitute a phrase shall contain at least one non-empty score atom.

---

[3]A famous exception with respect to the traditional academic rules (which prescribe scanning of phrases according to multiplies of four) is represented by the first two phrases of *St. Anthony's Chorale*, used by Brahms in his *Variations Op. 56*. In this case, phrases symmetrically proceed according to a scanning based on five measures (instead of four).

[4]We will discuss this problem in the following Chapter.

As an example, we might refer to the celebrated *incipit* of Beethoven's 9th Symphony, where the measures 2-5 played by the first violins, the violas and the double basses give rise to a phrase in this sense. Here, we suppose that the corresponding measures of all other instruments consist of score-holes ♣.

But what might be the interest of formalizing musical languages? As is well known, in the case of scientific theories, formalization is not aimed at providing some perfect languages that should substitute the "old rough" languages used by the scientific community. Formal languages are heavy and unreadable (if not accompanied by some translation-rules into the natural language). In a similar way, any attempt to substitute a traditional score with its formal version would be absolutely unreasonable! In both cases (science and music), the basic aim of formalizing languages is bringing into light some deep linguistic structures that represent significant invariants in a variety of different kinds of expressions and notation-systems. Identifying the elements that have a fundamental role in our information-encoding process represents the first step for any successful theoretic analysis.

CHAPTER 7

The relationship whole-parts in the semantics of music

How can we describe the relationship between a score and the class of all its (possible or real) interpretations? What do we exactly mean by *interpretation* of a musical score? We are now dealing with a quite critical concept, often discussed by musicians and musicologists, who have proposed different perspectives and solutions.

As already noticed, the sound-world is a typically *relational* world: the meaning of a single note, of a chord, of a musical phrase is always context-determined. There is no doubt that music requires a holistic and contextual semantics. Is it possible (and interesting) to look for a formal characterization of the intuitive concept of *musical interpretation* by using some abstract tools suggested by our quantum holistic semantics?

Consider a score **S** (say, the score of Beethoven's 9th Symphony), and let **FS** be a formal version of **S**. By abstracting from what happens in the concrete cases, let us first try and identify the elements that play a fundamental role in any possible interpretation of our score.

a) **The choice of musical phrases**

For any relevant score-fragment, any interpretation generally collects the signs occurring in that fragment by forming a few phrases that create a kind of dialogue in the framework of a characteristic "musical design". Consider a complex instrumental composition (say, a symphony by Brahms or by Mahler): the conductor (and the listener, as well) does not generally perceive distinct sounds that correspond to the different rows in the score. On the contrary, he (she) seems to "collect" them into some holistic patterns that behave like "individual voices". Of course, this does not forbid the (good) conductor to recognize possible mistakes of single performers in an analytical way.

Accordingly, from a formal point of view, the following assumption seems to be sensible: the first element of a given interpretation is a partition of the score into a number of musical phrases. We will call *score-covering* any set *Phr* of phrases that satisfies the following condition: each score-atom occurs exactly in one element of *Phr* (hence, *Phr* turns out to *cover* the whole score).[1] Needless to say, any score admits, in principle, a number of

[1] Notice that each score-atom is labelled by an index. Hence, different occurrences of one and the same note-name give rise to different atoms.

different coverings. As a consequence, one can reasonably assume that the choice of a covering Phr represents the first characteristic element of a given interpretation, giving rise to a significant bridge between musical syntax and semantics. Just in this sense the common musical jargon uses to refer to the *phrasing* of a particular interpreter.

b) **The tempo-choice**

Another element that represents a characteristic feature of any particular interpretation is the tempo-choice. As is well known, different performances of one and the same score may have quite different durations. All metronomic indications (contained in the score) generally represent only approximate prescriptions: should they be faithfully respected, the result would be a quite boring interpretation!

How can we describe, from an abstract point of view, the tempo-choice in the framework of a particular interpretation? We can assume that the second characteristic element of a given interpretation is a *temporal function* $Temp$ that assigns a time-interval to each column occurring in the score. Let

$$\mathbf{c}_r^j = \begin{pmatrix} A_{1j}^{[r]} \\ A_{2j}^{[r]} \\ \dots \\ A_{mj}^{[r]} \end{pmatrix}$$

be the j-th column of the r-th score-measure. We will write:

$$Temp : \mathbf{c}_r^j \mapsto \Delta t.$$

Of course, as happens in all experimental sciences, the lengths of such time-intervals are always determined up to a certain *error*: as a consequence, any $Temp(\mathbf{c}_r^j)$ is, in fact, a fuzzy interval. On this basis, the function $Temp$ automatically determines the duration of the performance of the whole score. From the musical point of view, $Temp$ realizes the performer's dynamic choices: any *accelerando*, *ritardando*, *rubato*, which often do not have any clear syntactical counterpart in the written score.

c) **The choice of musical meanings**

The choice of musical meanings certainly represents the most critical element for an abstract concept of musical interpretation. What do musical compositions mean? Does music have any content? As is well known, this question has been deeply discussed by musicians, musicologists and philosophers, who have proposed different answers. A basic problem seems to be the following: suppose we accept that music is associated to some *meanings*. Are these meanings always *internal* to music, or do they rather essentially refer to some *external worlds*? Whatever is our answer to this preliminary question, we are faced with the following problem: is it possible to define those particular objects that we call the *meanings* of a musical composition?

In fact, looking for simple definitions seems to be a somewhat naive attitude both in the case of music and of scientific theories. A more fruitful position, that has represented a winning trend for a number of contemporary semantic theories, refuses any research for general *explicit definitions*. Instead of trying to define the entities we are talking about, one attempts to describe their behaviour by creating some convenient structures, where our entities are supposed to live. On this basis *meanings* are identified with some abstract objects that belong to special worlds, often called *models* of the theories under investigation. Following this perspective, we will not try to answer the question "what are musical meanings?". We will only attempt to formally describe some characteristic properties thereof.

Our starting point is the following hypothesis: we assume the existence of a *universe* (indicated by $\mathbf{M}$) of ideal objects called *musical meanings*. The elements of $\mathbf{M}$ may possibly refer to some extra-musical worlds that are supposed to be distinguished from them. From an intuitive point of view, musical meanings can be thought of as a kind of *intensional* objects, that are quite similar to the intensional meanings associated to the expressions of our natural languages.[2] Generally, a musical meaning cannot be directly identified with a physical sound-event. One is rather dealing with an *idea* that permits us to create a particular sound-event. Just these ideas represent an important link between the score (which is a syntactical object) and a performance (which corresponds to a physical event). In a sense, they play the role of a kind of *invariants* (or *quasi-invariants*) with respect to the concrete performances that are realized in different times. It is not a chance that we use to speak, for instance, of "Claudio Abbado's interpretation" of Beethoven's 9th Symphony, without necessarily referring to a particular historical performance.

How is the universe of all musical meanings structured? It is natural to assume the following hypothesis: $\mathbf{M}$ includes a subuniverse $\mathbf{M}^{At}$ consisting of all *atomic meanings* that may represent the *intension* of a single note. As an example, in the case of a string quartet, the atomic meanings will be represented by all possible ideas of a single sound that can be performed by a violin or by a viola or by a cello. Apparently, when a quartet-performer plays a single note, he (she) selects a sound-idea from this atomic universe. We will suppose that the set $\mathbf{M}^{At}$ contains tho privileged elements: the *silence-meaning*, which corresponds to a pause-symbol, and an *empty meaning*, which represents the meaning of a score-hole $\clubsuit$.

As we know, single notes are combined giving rise to score-columns and to phrases (which are column-sequences possibly containing score-holes $\clubsuit$).

[2] The logical tradition uses to distinguish the *extensional* from the *intensional* meaning of a syntactically well-formed expression: while the *extension* represents the concrete reference, the *intension* corresponds to the conceptual content of the expression in question. For instance, in the case of a *predicate* (like "man", "planet", "triangle") the *extension* is the set of all objects that satisfy the given predicate (say, the set of all men, of all planets, of all triangles). The *intension*, instead, is the concept associated to our predicate: in other words, the idea that is generally understood by the persons who speak the language to which the predicate belongs.

In order to associate meanings to all molecular syntactical objects, we shall refer both to vertical and to horizontal combinations of atomic meanings. Let v be the *vertical complexity* (i.e. the number of rows) of our score: any sequence $(\mathbf{m}_1, \ldots, \mathbf{m}_v)$ of v atomic meanings will represent a possible meaning of a score-column. Such molecular meanings (expressing ideal chords) will be called *vertical meanings*. We will assume that $\mathbf{M}$ contains all possible sequences $(\mathbf{m}_1, \ldots, \mathbf{m}_v)$; in other words, $\mathbf{M}$ is supposed to be closed under the v-th cartesian power of $\mathbf{M}^{At}$.[3]

Any *analytical* semantic description would stop here, identifying the set of all vertical meanings (of complexity v) with $(\mathbf{M}^{At})^v$ (the v-th cartesian power of $\mathbf{M}^{At}$). In this sense, molecular meanings are represented as determined by their parts. As we have seen, quantum holistic semantics gives rise to a different situation: generally, the meaning of a global expression determines the meanings of its parts, and not the other way around! Furthermore, meanings may be entangled: hence our global understanding of a given expression might be more precise and sharp than the contextual pieces of information we have about its parts.

Let us try and apply these ideas to the semantics of music. Our basic hypothesis is the following: the universe of all vertical meanings (the ideal chords) is not realized by the v-th cartesian power of $\mathbf{M}^{At}$. Besides the *analytical meanings* (represented by all sequences $(\mathbf{m}_1, \ldots, \mathbf{m}_v)$), we assume the existence of *holistic meanings*, which are not determined by the combination of their parts. Following the example of quantum semantics, holistic meanings can be thought of as *superpositions* of analytical meanings, that ambiguously describe a variety of co-existent semantic situations. We will indicate by $\mathbf{M}^{\downarrow}$ the set of all possible vertical meanings, which may be either analytical or holistic; and we will use the symbol $\mathbf{m}^{\downarrow}$ for a generic element of $\mathbf{M}^{\downarrow}$.

As happens to composite quantum systems, any global vertical meaning $\mathbf{m}^{\downarrow}$ determines its parts, but generally not the other way around. We have seen that, in the case of quantum objects, one can define the *reduced state-function red*: if ρ is the state of a system consisting of n parts, then $red_i(\rho)$ represents the state of the i-th component of our system (for any i s.t. $1 \leq i \leq n$). It is reasonable to assume that a similar function also exists for musical meanings. Accordingly, we will call *vertical contextualization function* the following map:

$$cont_i : \mathbf{m}^{\downarrow} \mapsto \mathbf{m} \in \mathbf{M}^{At} \text{ (for any } 1 \leq i \leq v).$$

Hence, $cont_i(\mathbf{m}^{\downarrow})$ is the atomic meaning that represents the contextual meaning of the i-th part of $\mathbf{m}^{\downarrow}$. In the particular case where $\mathbf{m}^{\downarrow}$ is an

[3]We recall that the *v-cartesian power* of a set X is the set of all possible sequences consisting of v elements of X.

analytical meaning having the form $(\mathbf{m}_1, \ldots, \mathbf{m}_v)$, we shall require the following natural condition:

$$cont_i(\mathbf{m}^{\downarrow}) = \mathbf{m}_i.$$

In other words, the contextual meaning of the i-th part of $\mathbf{m}^{\downarrow}$ is just its i-th element. In fact, analytical meanings turn out to behave like *factorized states* in QT (states that can be represented as tensor products $\rho_1 \otimes \ldots \otimes \rho_n$ of the states ρ_i associated to the parts of a composite system).

As an example, let $\mathbf{m}^{\downarrow}$ be a vertical meaning corresponding to a C-major chord (which consists of the three notes C, E, G):

Suppose we are reasoning in the framework of a harmony-treatise, where we are not concerned with any interpretation-problem. In such a case we will naturally represent $\mathbf{m}^{\downarrow}$ as an analytical meaning identified by the sequence of three atomic meanings. Hence will have three vertical contextualization-functions such that:

$$cont_1(\mathbf{m}^{\downarrow}) \mapsto \mathbf{m}_1; \; cont_2(\mathbf{m}^{\downarrow}) \mapsto \mathbf{m}_2; \; cont_3(\mathbf{m}^{\downarrow}) \mapsto \mathbf{m}_3,$$

where $\mathbf{m}_1$, $\mathbf{m}_2$, $\mathbf{m}_3$ represent three atomic meanings corresponding to the three notes C, E, G, respectively.

A similar procedure can be also applied to horizontal combinations. In this way we obtain some abstract configurations that represent harmonic and melodic structures at the same time. Let o be the *horizontal complexity* of our score (which is determined by the number of the score-columns). Any sequence

$$(\mathbf{m}_1^{\downarrow}, \ldots, \mathbf{m}_r^{\downarrow}) \text{ (where } 1 \leq r \leq o),$$

represents a possible horizontal meaning, that may be the interpretation of a phrase whose (horizontal) complexity is r. Sequences of this kind will be called *horizontal analytical meanings* (of complexity r).

Of course, analytical meanings will not exhaust the universe of all possible horizontal meanings. As happens in the case of vertical meanings, we will have horizontal meanings that cannot be identified with sequences of vertical meanings. Such meanings, that are not elements of the r-th cartesian power of $(\mathbf{M}^{At})^v$, will be called *horizontal holistic meanings*. We will indicate by $\mathbf{M}^{\rightarrow}[r]$ the subset of $\mathbf{M}$ that consists of all (analytical and holistic) meanings of complexity r. The symbol $\mathbf{m}^{\rightarrow}[r]$ will represent a generic element of $\mathbf{M}^{\rightarrow}[r]$, while $\mathbf{m}$ will denote any musical meaning living in the universe $\mathbf{M}$. Apparently, all vertical meanings (which consist of a single column) are special examples of horizontal meanings whose complexity is 1.

As we have done in the vertical case, we will also assume *horizontal contextualization functions* that allow us "to go from the whole to the

FIGURE 1. The *incipit* of Beethoven's 9th Symphony

parts". Let $\mathbf{m}^{\rightarrow}[r]$ be a horizontal meaning: for any i (s.t. $1 \leq i \leq r$), $cont_i(\mathbf{m}^{\rightarrow}[r])$ will represent the i-th *contextual meaning* of $\mathbf{m}^{\rightarrow}[r]$. As expected, the contextualization-functions can be naturally generalized to all *molecular parts* (consisting of sequences of columns) of a given meaning $\mathbf{m}$. As an example, we might refer again to the *incipit* of Beethoven's 9th Symphony. Now we will consider the first five measures of the whole score.

Here the dominating phrase (E-A, A-E, E-A) is performed by the first violins, the violas and the double-basses. One is dealing with a musical idea that has been compared with a sudden, striking event. In this connection, one has sometimes evoked the image of a meteorite-fall, but we could also think of the origin of the Universe. This dominating phrase emerges from a kind of indistinct background: the empty fifth-chords realized by the horns and by the *tremolos* of the second violins and of the cellos. Needless to observe, any analytical analysis would be fully inadequate to describe such a semantic situation. One cannot avoid referring to a global musical meaning, while the different contextualization-functions determine different partial meanings for the component parts.

b) **The relationship phrases-meanings**

The set $\mathbf{M}$ of all musical meanings represents a kind of *virtual* universe: the score-interpretation shall select from this universe a particular meaning for each phrase that belongs to the score-covering Phr. Accordingly, any syntactical phrase P will be associated to a *semantic phrase* represented by a meaning $\mathbf{m}$ ($\in \mathbf{M}$) that shall respect the linguistic form of P.

We know that in the semantics of scientific theories, any model of a theory T associates to any expression of the language of T a meaning that lives in the system of objects created by the model in question. In a similar way, in our musical semantics it is natural to require the existence of a map *Real* that assigns to any phrase P of the score-covering Phr a convenient meaning:

$$Real : P \ \mapsto \ \mathbf{m} \in \mathbf{M},$$

where the phrase P and $Real(P)$ are supposed to have the same horizontal complexity r.

On this basis, the contextualization-functions, applied to $Real(P)$, will determine a contextual meaning for all sub-phrases of P. We require that the realization-function *Real* satisfies the following (natural) condition: the contextual meaning of any occurrence in P of the score-hole $\clubsuit$ is always the empty meaning.

To what extent shall the map *Real* preserve all pieces of information that are written in the score? In principle, any "faithful interpretation" should respect *all* score-prescriptions. We know however that this is not generally the case in real performances, both for technical mistakes and for interpretation-choices of the performers. How shall we formally describe the "faithfulness-degree" of a given interpretation? As happens in the case of experimental sciences, we can require that *Real* preserves the score-prescriptions *up to a certain accuracy*, that can be conventionally represented by a numerical parameter ε_{Real} (which can be easily defined). Of course, ε_{Real} will represent one of the characteristic features of the interpretation in question.

The conditions we have required guarantee the contextual behaviour of *Real*: phrases that occur in different parts of the score generally receive different contextual meanings. This reflects what happens in the case of real performances, where repeated musical phrases are generally interpreted according to different modalities[4] .

e) **The choice of extra-musical meanings**

[4] As observed by the conductor Piero Bellugi, the "copy and paste operation" cannot exist in music.

So far we have considered musical meanings as a kind of autonomous ideal objects that may possibly refer to some extra-musical worlds (which are sharply distinguished from them). We know however that evoking some extra-musical events (like emotions, feelings, descriptions, ...) represents, in many cases, an important characteristic property of a given interpretation. And referring to extra-musical situations turns out to be an essential component of interpretations, whenever the score includes a text (as happens in the case of lyric operas and of *Lieder*).

How can we find a formal counterpart for such correlations between musical and non-musical events? We can assume that the last characteristic element of a given interpretation is represented by a map Wor that assigns to any phrase P of the score-covering Phr a world w consisting of non-musical meanings:

$$Wor: \ P \ \mapsto \ w.$$

It seems reasonable to admit that $Wor(P)$ may be the empty set. In fact, there are musical compositions (or parts thereof) whose meanings are completely internal to music. As an example, we might think of many Bach's compositions (like the *Goldberg Variations*).

What can be said about the values of the map Wor? We should try and represent the vague, ambiguous and subjective features that seem to characterize all feelings and all concrete situations that are evoked by music. Classical semantics could not certainly be used to this aim. However, we can have recourse to a special kind of *fuzzy possible world semantics*, where generally all objects and all relations are supposed to have an ambiguous behaviour. In this framework, $Wor(P)$ can be identified with a particular *fuzzy possible world*. Needless to say, the creation of the fuzzy worlds to be associated to the different score-phrases, represents a characteristic choice of any interpreter.

Summing up, any interpretation $\mathcal{I}$ of a formal score **FS** can be now represented as an abstract structure whose form is the following:

$$\mathcal{I} = (Phr, Temp, \mathbf{M}, Real, Wor),$$

where:

- Phr is a score-covering of **FS**;
- $Temp$ is the function that assigns to any score-column of **FS** a time-interval;
- **M** represents the universe of all musical meanings, which includes the sets of all atomic, of all vertical and of all horizontal meanings. For any meaning **m** (of **M**), the (vertical and horizontal) contextualization-functions assign contextual meanings to all parts of **m**.
- $Real$ is the realization map that assigns a semantic phrase (an element of **M**) to any syntactical phrase P of Phr, by preserving the score-prescriptions (contained in P) up to a certain accuracy ε_{Real}.

- Wor is the map that assigns to any phrase P of Phr a (possibly empty) fuzzy world w, which represents the extra-musical meaning of P.

We require that $\mathcal{I}$ satisfies all the conditions we have illustrated above.

Following the example of (formalized) scientific theories, we can now identify any musical composition with a pair $(\mathbf{FS}, \mathbf{K})$ consisting of a formal score **FS** and of the class **K** of all possible interpretations $\mathcal{I}$ of **FS**. On this basis, the *history* of a given musical composition can be reconstructed as a kind of "journey" through the class **K**. Since the set of all historical journeys is *de facto* finite, while **K** is in principle infinite, one can easily understand the abstract reason why musical compositions are essentially *unfinished and open works*.

CHAPTER 8

The *Lied*-form and Mignon's enigma

Lieder represent a musical form that can be naturally analyzed in the framework of our quantum holistic semantics. As is well known, the characteristic property of *Lieder* is a "co-existence" between three different worlds: one created by a poem (which is, in many cases, a classical literary work, with an autonomous life and history[1]), a second one consisting in the melodic line of the singing, and - last, but not least - the instrumental part, whose complexity may be, to a certain extent, independent both of the text and of the vocal melody.

A crucial question (often discussed by musicologists) concerns the relationship existing between the text and the music of a *Lied*. Apparently, each musical realization transforms a given text into a completely new *global semantic object*, where the original text represents, however, an essential component.

Arnold Schönberg has written:

> *A few years ago I was deeply ashamed when I discovered in several Schubert songs, well-known to me, that I had absolutely no idea what was going on in the poems on which they were based. But when I had read the poems it became clear to me that I had gained absolutely nothing for the understanding of the songs thereby, since the poems did not make necessary for me to change my conception of the musical interpretation in the slightest degree. On the contrary, it appeared that, without knowing the poem, I had grasped the content, the real content, perhaps even more profoundly than if I had clung to the surface of the mere thoughts expressed in words.*[2]

At first sight this assertion may appear somewhat paradoxical. However, one can reasonably guess that Schönberg is here referring just to the global musical meanings of *Lieder*, that somehow absorb and renew all the partial meanings of the original literary texts.

[1] In some cases poets themselves have conceived their *Lieder* with peculiar attention to the inner musical structure of the verses.

[2] A. Schönberg, "Das Verhältniss zum Text", *Der Blaue Reiter*, München, 1912. English translation of the author in *Style and Idea*, Philosophical Library, New York, 1950.

FIGURE 1. Mignon and the Harpist

There is an interesting test we can perform. What happens when one and the same poem has been set to music by different composers (in different times)? We might refer to a number of classical examples of the *Lied*-history, as the celebrated *Lieder* sung by *Mignon* and the *Harpist* in Goethe's *Wilhelm Meisters Lehrjahre* (Fig. 1).

Let us consider, for instance, the famous poem "Kennst du das Land" ("Know'st thou the land") and let us try and compare the musical realizations by Schubert (1815) and by Schumann (1849).

In Goethe's novel Mignon's character is deeply enigmatic: she is represented at the same time as a child (often called "das Kind") and as a young woman who is passionately in love. Her dominating feeling is a sense of nostalgic desire (*Sehnsucht*) for a mysterious past that she does not really remember, although she has sworn that she would never reveal it. In "Kennst du das Land" Mignon tries to reconcile her feeling of love and homesickness with her wish to return to the land of her vague and obscure recollections.

Lied der Mignon
Kennst du das Land, wo die Zitronen blühn,
im dunklen Laub die Goldorangen glühn,
ein sanfter Wind vom blauen Himmel weht,
die Myrte still und hoch der Lorbeer steht?
Kennst du es wohl?
Dahin, dahin möcht' ich mit dir, o mein Geliebter ziehn!

Kennst du das Haus, auf Säulen ruht sein Dach,
es glänzt der Saal, es schimmert das Gemach,
und Marmorbilder stehen und sehn mich an:
was hat man dir, du armes Kind, getan?
Kennst du es wohl?
Dahin, dahin möcht' ich mit dir, o mein Beschützer ziehn!

Kennst du den Berg und seinen Wolkensteg?
Das Maultier sucht im Nebel seinen Weg,
in Höhlen wohnt der Drachen alte Brut,
es stürzt der Fels und über ihn die Flut:
kennst du ihn wohl?
Dahin! Dahin geht unser Weg; o Vater lass uns ziehn![3]

Schubert's and Schumann's intonations of this *Lied* are deeply different, although both of them express the enigmatic and mysterious features of Mignon's character.[4] In Schubert's composition one can feel something "consoling": the *incipit* (in a major tonality) may sound as a quiet lullaby ("Kennst du das Land, wo die Zitronen blühn, im dunklen Laub die Goldorangen glühn"). Soon after, the verses "Ein sanfter Wind vom blauen

[3]Know'st thou the land where the lemons flower, / 'Mid shady leaves the golden oranges glow, / A gentle breeze blows from a blue heaven, / The tranquil myrtle and the bay trees thrive?/ Know'st thou it? / Thither, oh my love, / I long to go with thee! / Know'st thou the house? On pillars rests its roof, / The hall is bright, the chambers dimly glow, / And statues stand and gaze down on me; / Poor child, what have they done to thee? / Know'st thou it? /Thither, oh my protector, / I long to go with thee! / Know'st thou the mountain and its cloudy path? / In the mist a mule picks its way, / In caves the dragon's ancient brood doth dwell; / The rocks will tumble, and after them the flood. / Know'st thou it?/ Our way leads thither,/ Father, let us go!

[4]The scores of both *Lieder* are in the Appendix.

Himmel weht, die Myrte still und hoch der Lorbeer steht?" seem to describe some joyous images, also expressed by the bright triplets of the piano-accompaniment. Doubts and a slight anxiety are only allusively expressed by the dissonant chords lying under the question "Kennst du es wohl?". Should we only know Schubert's *Lied* (ignoring Goethe's novel) we might guess that Mignon's and Wilhelm's story is compatible with a "happy end". In particular, the finale of the *Lied* (in major), where the word "dahin" is repeated four times, seems to suggest an optimistic impetus, a kind of *Streben* towards a goal that cannot be reached, although some dissonant elements in the precipitous instrumental part could cast a shadow of doubt.

Schumann's *Lied*, instead, is completely dominated by a sense of pain and of anguish that seem to anticipate Mignon's tragic epilogue. The melody is somehow "folded" upon itself, tending mostly to descending lines: this is emphasized by the choice of a minor tonality and by an insisting sequence of descending minor-second intervals. Paradoxically, even the verse "Und hoch der Lorbeer steht" is musically realized by a descending melodic interval. At the same time, the piano performs a number of diminished and dissonant chords, starting from the introduction that immediately creates a strong emotional "implosion". While Schubert's vocal *incipit* is characterized by a positive sign, Schumann's "Kennst du das Land" contains an anxious question-mark, realized by an ascending fourth-interval that interrupts the joint-degree sequence of the first measures. Soon after, his "Ein sanfter Wind ..." does not suggest any bright image. What is communicated is a kind of "storm" of anguished feelings, expressed by a sudden extension of the voice (spreading over a ninth-interval) and by anxious chromatic lines in the chord-triplets of the piano accompaniment. Unlike Schubert, Schumann does not end with the word "dahin", but instead with the verse "O Vater, lass uns ziehn!", which expresses a painful pray. The conclusion is committed to the piano: a question-mark evoked by means of an ascending and unresolved tonic-dominant interval, which leaves us somehow suspended above an abyss.

Of course, it would be unreasonable to ask: which *Lied* is more faithful to Goethe's poem? In a sense, each musical realization creates a new poem that is a part of a *global semantic object*. Is it possible to go back and obtain again the original poem in the *Lied*-context? In the holistic semantics-framework such an operation can be described in terms of the *contextualization-functions* (illustrated in the previous chapter). Anyway, one should sharply distinguish the *contextual meaning* of the poem (which is an internal component of the *Lied*) from Goethe's text. If we want to discover again the original poem (quite independently of the musical realization), we shall perform a kind of *analytical reading* of the literary text. One is dealing with an operation that seems to have some characteristic features of quantum-measurements.

The holistic aspects of the relationship music-text represent perhaps one of the reasons that might explain some psychological and cognitive behaviours of singers. Let us just consider how a singer may perfectly know a *Lied* or an opera-aria by heart, but, in spite of this, declaiming the text separately from the music often turns out to be difficult and artificial for that same singer. And it is quite improbable that a singer confuses two different *Lieder*, only because they are based on the same text!

The contextual behaviour of our "musical brain" seems to be confirmed by some neuroscience-experiments, that have been realized by means of *brain-imaging* techniques. For instance, an interesting case is concerned with a pianist who has been analyzed in two different musical situations. In the first one he was playing a simple F major-scale; while in the second situation he played the same scale in the context of a Bach's composition (the third movement - *Presto* - of *Concerto italiano BWV 971*). The result was that in the two different situations, two different brain-regions of the performer had been activated. This seems to confirm that one and the same musical input (a scale-performance) is perceived and elaborated by our brain according to different modalities, which are context-dependent.[5]

Some interesting remarks about the holistic aspects of music can be also found in Penrose's work. As an example, let us quote the following significant passage from the last Chapter of *The Emperors's New Mind*:

> *Listen to the quadruple fugue in the final part of J.S. Bach's* Art of Fugue. *No-one with a feeling for Bach's music can help being moved as the music stops after ten minutes of performance, just after the third theme enters. The composition as a whole still seems somehow to be 'there', but now it has faded from us in an instant. Bach died before he was able to complete the work, and his musical score simply stops at that point, with no written indication as to how he intended to continue. Yet it starts with such an assurance and total mastery that one cannot imagine that Bach did not hold the essential of the entire composition in his head at that time. Would he have needed to play it over to himself in its entirety in his mind, at the normal pace of a performance, trying again and again, and yet again, as various different improvements came to him? I cannot imagine that it was done this way. Like Mozart, he must somehow have been able to conceive the work in its entirety, with intricate complication and artistry that fugal writing demands, all conjured up together. Yet, the temporal quality of music is one of its essential ingredients. How is it that music can remain music if it is not being performed in 'real time'?*[6]

[5]See L.M. Parsons, J. Sergent, D.A. Hodges, P.T. Fox,"The brain basis of piano performance", *Neuropsychologia* **43 (2)**, pp. 199-215, 2005. We thank the neuroscientist Sandro Sorbi who has informed us about this interesting experiment.

[6]R. Penrose, *The Emperors's New Mind*, Oxford University Press, Oxford, 1989, p. 576.

CHAPTER 9

Quantum-like superpositions in music

1. Ambiguity and superpositions

We have seen that the quantum formalism is characterized by a systematic use of *superpositions of states*, which allows a kind of "co-existence" of different antagonistic properties. A paradigmatic case is the two slit-experiment, where particles seem to follow two different trajectories at the same time. Interestingly enough, such a strange behaviour gives rise to some useful applications in the framework of quantum computation, where atomic pieces of information (*qubits*) are represented as *superpositions* of the two classical truth-values, the *Truth* and the *Falsity*.

Semantic objects that seem to behave like quantum superpositions can be naturally recognized also in the framework of musical structures. An example is represented by the *non-analytical musical meanings* (described in Chapter 7). These can be generally reconstructed as a kind of ambiguous superpositions of some analytical meanings, according to the rules of our abstract holistic semantics. Consider again the piano-conclusion of Schumann's *Lied* "Kennst du das Land". One is dealing with a quite simple musical phrase, which seems to suggests a question that is somehow suspended in an empty space:

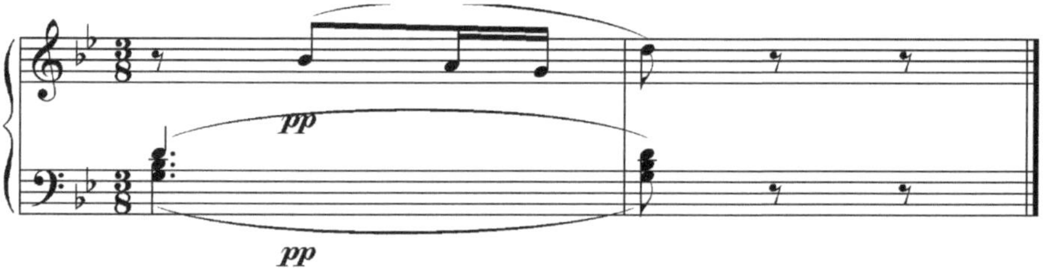

The meaning of our phrase is typically contextual with respect to the structure of the whole *Lied*. Hence, it cannot be simply identified with the horizontal combination of four degrees of the G minor scale (third, second, first, fifth), which might appear, in a sense, trivial. How shall we read the semantic phrase, corresponding to such note-sequence, as a quantum-like superposition? Any reference to a classical *set* consisting of several instances of note-sequences (B flat, A, G, D) would be quite misleading. In fact, quantum superpositions do not represent sets, but rather relatively simple ideal objects that *virtually* refer to infinitely many possible instances. Only the *collapse of the wave-function* induces a transformation of an *uncertain*

superposition into a *certain* state. In the same way, the final semantic phrase in Schumann's *Lied* can be regarded as an essentially vague and ambiguous musical idea, which is compatible with many different analytical realizations. The semantic situation is quite similar to what happens with the last verse of Leopardi's *Infinity* (*And drowning in this sea is sweet to me*). Also in this case the contextual meanings of the expressions "drowning", "sweet", "sea" can be represented as superpositions that virtually and vaguely refer to several different meanings that are currently associated to such words.

Superpositions play an important role not only in the case of *pure musical meanings* (the ideal objects that live in the universe **M**): also the *vague worlds*, evoked by music (we have analyzed by means of the abstract function Wor) can be reconstructed as superposition-states. In fact, many descriptions and emotions that are evoked by music are often based on antagonistic elements that seem to co-exist. The pair *true- false* may assume, in different situations, different "disguises": *serious-humorous*, *tragic-comic*, *good-evil*, and so on.

Musical examples that clearly realize such contrasts can be easily recognized in Mozart's operas. Think, for instance, of the famous duet "Che soave zeffiretto" between Susanna and the Countess in the opera *Le nozze di Figaro* (*The marriage of Figaro*). This represents a paradigmatic case where *true* and *false*, *ironic* and *melancholic* are mixed in an intriguing and ambiguous way. Both characters are aware that they are organizing a joke and a deception against the Count; in spite of this, the *false* love-rendezvous "nel boschetto" (in the thicket) is musically evoked according to a romantic *rêverie* that transforms a *falsity* into a *true* and *authentic* event, where any fiction seems to disappear.[1] Needless to observe how effects of this kind are quite impossible in the framework of theater-situations that do not make use of music.

There is another opera by Mozart, where the superpositions *true-false*, *ironic-melancholic* seem to represent the essential key of the whole composition. In *Così fan tutte* (*All Women do the Same*) all feelings appear *false* and *genuine* at the same time. The opera's structure is based on complicated symmetry-games. At the same time, music seems to suggest the unavoidability of love's desire (whose object might remain ambiguous and indefinite) and the impossibility of realizing love in a full and authentic way. And in many parts of the opera the dominating feeling is just a generalized melancholy, which transforms something that is basically false into something else that appears true.[2]

Wagner's *Der Ring des Nibelungen* (*The Nibelung's Ring*) represents another characteristic example where musical and narrative superpositions seem to be entangled according to modalities that are emphasized by the

[1]Any performance of the duet which only stresses the ironic aspect of the situation clearly represents a "betrayal" of the score.

[2]As an example, we might mention the quintetto "Di scrivermi ogni giorno", the terzetto "Soave sia il vento" or the aria "Un' aura amorosa".

Leitmotiv-technique. Almost all main characters of the *Ring* are deeply ambiguous, representing a kind of "node of contradictions". Wotan, the sad god, is constantly torn by antagonistic ideas and feelings: legality and fraud, love's desire and renouncement of love, construction and destruction. Also Fricka, his wife, has a contradictory behaviour. In *Das Rheingold* (*The Rhinegold*) she appears as the most human of all gods: she strongly defends her sister Freia against Wotan, who had cynically decided to sacrifice the goddess of youth to the Giants as payment for their work in constructing Walhalla. In *Die Walküre* (*The Valkyrie*), instead, the roles are completely inverted: the most human is Wotan, who defends the guilty love of Siegmund and Sieglinde. But Fricka objects, evoking rigid and ideological principles of marriage-legality. And finally the Lord of the gods surrenders to his wife's will, parting for ever from Brünnhilde, his favourite daughter who had disobeyed his command in order to save Siegmund's life. Finally, in *Götterdämmerung* (*The Twilight of the Gods*) all contradictions explode: the heroic Valkyrie, who had defended the reasons of love before knowing love, is transformed into a kind of "bourgeoise" wife, who is patiently waiting at home for her husband's return. And when she discovers Siegfried's treason, she takes revenge determining his death. Sigfried himself is only superficially a "pure hero" ("ein reiner Held"). In fact, he behaves as if he were a superposition of a "superman" and of a "naive stupid boy". The potion, that the wicked Hagen had induced him to drink, seems to bring into light a deep aspect of his: the nature of a trivial seducer. Contradictions are only solved at the end of the *Ring*, in the final catastrophe where fire destroys Walhalla and all gods.

Other superposition-cases concern directly the *musical forms*. A significant example is the last movement of Mozart's "Jupiter" Symphony, which can be analyzed as a superposition of two alternative musical forms: the *sonata form* and the *fugue*. We will discuss this case in the next Chapter.

2. Florestan and Eusebius

There is a great composer who experienced the superposition of different psycho-musical states in an extreme way: it is Schumann, who tried to present his musical personality as a "combination" of two opposing characters, *Florestan* and *Eusebius*. The theme of the split personality (dear to early Romantic authors such as Jean-Paul Richter and Ernst Theodor Amadeus Hoffmann) constitutes a veritable principle of construction in Schumann, determining the kaleidoscopic expressive physiognomy and, at the same time, the deep internal unity of many of his compositions.

Who exactly were Florestan and Eusebius? Schumann's theoretical and critical articles, published in the review *Neue Zeitschrift für Musik* (founded by Schumann and edited by him from 1834 to 1845), are often signed by either Florestan or Eusebius and on occasions by them both. In general the opinions and ideas proposed by the two characters do not appear to be antagonistic or contradictory; indeed they often represent complementary points of view, put forward as equally true. To continue using a quantum language we might say that the co-existence between Florestan and Eusebius could be seen as a sort of superposition of distinct *non-orthogonal* states.

Through these imaginary characters, Schumann wanted to describe the various aspects of his human and musical "ego": Eusebius' lyrical and dreamy melancholy and the ardent (sometimes heroic) passionate enthusiasm of Florestan alternate in our imagination and reassemble as the smiling and superior sagacity of *Meister Raro*, who marks Schumann's work with doctrinal perfection and musical practice.

This can be exemplified by reference to the famous *Phantasiestücke Op. 12* (for piano), where the presence of Florestan and of Eusebius are clearly characterised. With *Des Abends* (*In the Evening*), the composition opens under Eusebius' influence: the structure of the piece is very simple, basing itself on the unravelling of an extremely delicate melodic line, stretched in long singing curves, made more limpid by a rhythmic pattern that attempts to escape the regularity of the metric accents. A rarefied and enchanted dimension of timbre is created, in an atmosphere suspended in the sound of Nature's mysterious poetry.

With the intense burst of the incisive and passionate theme of *Aufschwung* (*Soaring*) it is Florestan who asserts himself: after the exposition of the main theme, the second section is made up of the pursuit of a trochaic motif against an incessant flowing of sixteenth-notes in the middle register. The central part is formed of a sort of *chorale*, whose serene melodiousness soon leads to a sombre sequence of low chords, laid on a mysterious swirl of bass notes that prepares for the return to the first episode, concluding with two eloquent and vigorous chords.

Eusebius returns in *Warum?* (*Why?*): the atmosphere is once more intimate and surrounding an unresolved question, entrusted to the resumption in various voices of a small thematic cell that is gradually enriched with new harmonic and expressive emphases. The insistent and mournful repetition exceeds the limits of personal passion to become a part of a cosmic yearning for the unreachable infinity of the *All*, while its last appearance leaves the question suspended and unanswered. It is significant that the musical structure of the finale of *Warum?* (with the characteristic hanging on the fifth degree of the scale) is very similar to the piano conclusion in Schumann's version of "Kennst du das Land" ("Knowest Thou the Land?"), composed twelve years later.

In *Grillen* (*Whims*) Schumann seems to want to resolve positively the doubts in *Warum?* through a daring and slightly burlesque bonhomie, that is perhaps a glimpse of the presence of *Meister Raro*: it is in ternary tempo and of the two motifs in the first section the first one is really a variant of the theme of *Warum?*, while the second one is characterised by a singing and soaring lack of constraint that is almost waltz-like. In the central part, with a swell of syncopated chords, the mood goes from a melancholy sweetness to a rebellious and contradictory whim of fantasy that reigns supreme in the emotional aesthetics of the piece.

It seems that Schumann thought a lot of *In der Nacht* (*In the Night*), preferring it to all the other pieces of the series and recommending it to Clara Wieck, the great pianist who would become his wife. In the first part there is the insistent repetition of a two-note motif, that sounds like a sorrowful cry upon the lively, whirling arpeggio-motion of the low notes, while the melodic line is notched with brief resonant passages that seem to flash like beams of light in a dark and trembling atmosphere. The central section, instead, is characterised by a tender, lyrical outpouring with a more melodic motif in a slower tempo, immediately followed by a tumultuous transitional episode, run through with dark eighth notes, that rushes (in the *recapitulation*) with a violent and dramatic conclusion.

Eusebius appears again in *Fabel* (*Fable*), in which a slow phrase alternates with a playful motif, while the central part is made up of a more ample and animated episode; in the recapitulation the sequence of the two themes of the initial part is reversed, ending the passage in the stillness of the tender narrative fragment with which it started.

In *Traumes Wirren* (*Restless Dreams*) the whole composition is characterized by a continual swell in sixteenth-notes, which gives a sense of febrile anxiety to the sparkling and slightly hallucinatory merriment of the theme, only briefly interrupted by a secondary idea, made up of a linking together of mysteriously tranquil chords that seem to hang there and then rush once more into the troubled frenzy of the first section.

The epilogue of the long, fantastic adventure is marked by Meister Raro's smile: the final composition, *Ende vom Lied* (*End of the Song*), sets out with a serene and almost self-deprecatory motif (Schumann's dynamic indication is "Mit gutem Humor", "with good humour"), briefly developed and alternated with a more joyful passage, whose thematic starting-point is taken from the first motif of *Traumes Wirren*. An unpredictable *Coda* follows where a sweet effusion of melancholy overlays the initial optimism. In March 1838 Schumann wrote about this piece to Clara: "At the end I wanted to resolve everything in a happy 'marriage', but in the final bars the painful desire for you returned and now it sounds like a pealing of wedding bells and a death knell put together". Thus the ambiguous love-death superposition is evoked here too.

Florestan and Eusebius appear reconciled (or even unified) in the case of many Schumann's compositions. A very significant example is represented by the enigmatic *Geistervariationen über den lezten Gedanken WoO24* (*Variations of the spirits about the last thought*) for piano, composed in 1854, during a dramatic period of Schumann's life (when he attempted to commit suicide in the Rhine and was afterwards sheltered in Endenich's asylum). The composition consists of five variations on a very simple Theme (in *E* flat major) characterized by an essential tonal stability. The melodic line often delays on the measures' weak beats by means of syncopated and dotted notes, creating in this way an atmosphere that evokes a psychological suspension. All five variations have the same structure of the Theme (two parts, where the second one has a *refrain*), consisting of the same number of measures (with the exception of the last variation, where the finale has two further measures). The Theme remains almost always identical to itself, being repeated in the different variations as a kind of "fixed idea". Only the fourth variation has a tonality-change (in G minor); but the melodic line is not modified. The enigmatic character of the composition is compatible with many different interpretations. The Theme, repeated in an obsessive way, seems to evoke an absolute sorrow or also an objective fate that cannot be escaped. In the variations, the Themes's "fixed idea" (whose tempo is 2/4) is often entangled with triplets which seem to express a subjective reaction to something that appears inexorable. In the third variation, for instance, some high notes may suggest the idea of a crying. As happens in other Schumann's compositions (think, for instance, of the piano-conclusion in the cycle *Dichterliebe*), also in this case sorrow seems to be accompanied by a kind of acceptance of the fate, with a consoling attitude (which can be mainly recognized in the finale of the last variation). The ambiguity of the composition leaves to the performer a great interpretation-freedom. From a semantic point of view, we can refer to different superpositions in some abstract spaces, where the *basis-elements* codify, in some way, painful or consoling feelings.

CHAPTER 10

A superposition of sonata form and of fugue in Mozart's "Jupiter" Symphony

Compositions, whose formal structure is compatible with different reading-keys can be frequently found either in the symphonic or in the chamber music of the 18th and of the 19th Centuries. A significant example is Haydn's Quartet *Op. 76 No. 2*. Here the *Minuetto*, although preserving the rhythmic scansion of the dance, the ternary tempo and the characteristic colour of its historical origin, follows at the same time the rigid contrapuntual laws of a *Canon at the octave*.

The two forms (the *Minuetto* and the *Canon*) give rise to a superposition, where the autonomous identity of both forms is preserved. As a result, different listening-attitudes are possible: the attention may be directed either to the most evident form (the *Minuetto*) or to the counterpoint-procedure of the canon, where the parts undergo a phase displacement, while repeated melodies are interlaced. Actually, an ideal listening should simultaneously realize the two superposed forms: the dance and the canon. This probably corresponds to the authentic composer's intention: in a sense, only a double perception of the composition allows us to solve its ambiguity.

Another significant (and longer) example is represented by Franck's *Symphonic Variations*. The composition perfectly respects the basic principles of the form *Theme and Variations* (in this case, one is dealing with variations on two themes). At the same time, by analyzing the global structure of the work, we can discover that the composition is also articulated as a *Concerto-form* in three movements, where the second movement is an *Adagio*. Further examples of variations that include fully developed canons (which preserve, at the same time, the harmonic structure of the original theme) can be found in the monumental *Goldberg Variations* by Bach.

But the most genial case of an ambiguous superposition of forms is, without any doubt, the last movement of Mozart's Symphony *K. 551* ("Jupiter"). What a first listening perceives is a continuous springing up of motifs that are interlaced and superposed in a joyful rush towards the Symphony's solution. *Prima facie* the finale appears as a typical brilliant example of the Baroque literature; a careful analysis, however, shows the possibility of a double reading-key. The first one brings into light the structure of a sonata form, consisting of an *Exposition*, of a *Development*, of a *Recapitulation* and of a *Coda*. The second one gives rise to a more complex description. "This is not a fugue!" - have asserted some commentators. However, there are convincing reasons that seem to confirm the claim "This is more than a fugue!" In fact, the most severe procedures of a *fugato-counterpoint* clearly appear in the sonata-form structure with a surprising frequency: imitations, short canons, inverted motifs, contractions of the main theme, lightly flowing dilatations, whose aim seems to be hiding the composer's labour. By analyzing the thematic materials we can easily discover that themes and motifs have a common property: all of them are short, concise and incisive, determining the same acoustical weight within the general context. All this has a precise role in the fugato-treatment.

In the last measures of the movement's *Coda* the exposition of a four-subject fugue suddenly appears: all themes and motifs, elaborated during the exposition and the development of the sonata form, are here recapitulated and condensed into a unique block, which has a high polyphonic value. This represents the beginning of a perfect fugue, that a smiling Mozart immediately interrupts by means of a sequence of traditional conclusive chords. In order to fully grasp the second reading-key of the movement (the fugue-structure), a well-trained musical ear and a good acquaintance of musical events are no longer sufficient. All episodes are developed in a kind of thick lattice; as a consequence, the listener does not have enough time to perceive the ensemble of all procedures: superpositions, quick imitations, canons in several parts. Actually, a full understanding of the composition also requires another sense-organ: the view, that allows us to loose and to solve the form-ambiguity. The finale of the "Jupiter" Symphony should be read and listened at the same time!

Now it will be expedient to recall the basic formal properties of the sonata form and of the fugue (for the readers who are not expert of music). The structure of the *sonata form* is generally based on two main themes that alternatively recur and are continuously transformed, giving rise in this way to a well-identifiable melodic and rhythmic material. The two themes, that seem to behave like two logical ideas or also like two characters in a novel, often have quite different features; as a consequence the contraposition thereof turns out to be highly significant. Besides the two-theme construction, another characteristic property of the sonata form is the three-partition of the structure consisting of three phases: the *Exposition*, the *Development* and the *Recapitulation*. The aim of the Exposition is presenting the basic musical ideas (the themes). The Development realizes different possible transformations and consequences of the themes. Finally, the Recapitulation sums up all ideas as a conclusion of the musical journey. Both the Exposition and the Recapitulation include some passages (called *Transitions*), whose aim is creating a link between the two themes, soon after the presentation of the first theme; the procedure is quite similar to what happens during an argumentation, when one moves from some concepts to some new ones. At the end of the Recapitulation, one can find a *Coda*, whose aim is asserting again and more clearly the conclusion of the whole construction. Generally, the sonata form represents the basic structure of the first movement of symphonic and of chamber-music compositions (sonatas, symphonies, quartets, and so on). Sometimes, also other movements, whose tempo is usually *Allegro*, may have the same structure.

Fugue is a contrapuntal form. Thus, the musical construction is developed by means of superpositions of melodic lines: voices or instruments (generally called *parts*), which are stratified one above another along horizontal lines. Unlike the sonata form, the fugue-structure is generally based on a single musical idea, characterized by clear melodic and rhythmic properties. As a consequence, each appearance (*entrance*) of the idea in question can be immediately identified during the listening. Generally, one can say that fugue represents a *mono-thematic* form, although there are some complex examples of fugue with more than one subject.

The fugue-structure is essentially constituted by three phases: the *Exposition*, the *Divertimenti* and the *Stretti*. Let us refer to a fugue in four parts. In such a case the subject is stated by the first part and is immediately repeated by the second part, which asserts the *answer*, while the first part continues by proposing the *countersubject*. The answer represents a new statement of the subject on a different degree of the scale (whose distance from the subject is generally either a fifth or a fourth interval). Once completed the exposition of the subject and of the countersubject in the first two parts, the third part in turn states the subject in the main tonality; then, the fourth part answers, while all other parts continue developing a free counterpoint. After the exposition, the fugue goes on presenting again the subject, which is stated by different voices in different tonalities. Between one and another entrance of the subject, there are some passages called *Divertimenti*[1]. Soon after, the fugue continues alternating entrances

[1]The term *Divertimenti* is used in the sense of its etymological meaning, deriving from the Latin verb *devertere* (to divert).

and divertimenti in different tonalities. Finally, the conclusion is characterized by the *Stretti*, where the musical web becomes thicker and thicker, because the subject is stated more and more closely by all voices.

Roughly, the structure of the exposition of a fugue in four parts (I, II, III, IV) can be schematized as follows:

$$I: \quad SCFF$$

$$II: \quad AFF$$

$$III: \quad SC$$

$$IV: \quad A,$$

where:
S = *Subject*; C = *Countersubject*; A = *Answer*; F = *Free counterpoint.*

Let us now come back to the last movement of the "Jupiter" Symphony. How are the two structures (the sonata form and the fugue) superposed in this composition? First of all, let us isolate the two main themes of the sonata form:

First Theme

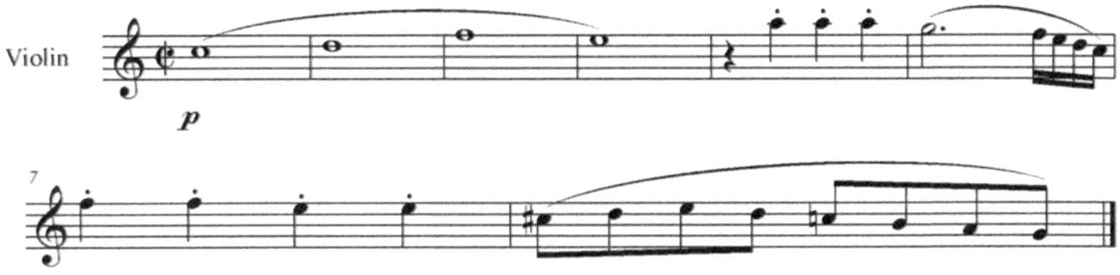

Second Theme

Besides the two main themes (representing the basic ingredients of the sonata form) two shorter themes (conventionally labelled as *Motif A* and *Motif B*) play here an important role.

Motif A

Motif B

These four themes seem to behave like some characters (or masks) in the framework of a unique theatrical performance.

Should we want to "read" the fourth movement of the "Jupiter" *Symphony* as a sonata form, we could easily recognize the sequence consisting of the exposition, the development, the recapitulation and the coda. In this structure the *First Theme* and the *Second Theme* play their usual role, while *Motif A* and *Motif B* act in the transition-passages. The "fugue-reading", instead, leads us to isolate several *fugatos*; the climax is given by the final *great fugato*, which appears in the *great coda* of the sonata form. This corresponds to the exposition of a fugue with four subjects, represented by the *First Theme*, the *Second Theme*, *Motif A* and *Motif B*, respectively.

The "double structure" of the movement can be synthetically described by the following scheme.

EXPOSITION

1. *First Theme*
2. *Motif A*
3. New exposition of the *First Theme* as a fugato
4. *Motif B*
5. *Motif A* with canonic imitations
6. *Second Theme*
7. *Motif A* and *Motif B* superposed
8. Second Theme repeated
9. *Motif A* and *Motif B* superposed. Imitations of the two motifs
10. Canonic imitations with the *incipit* of the *Second Theme*
11. *Second Theme* in a close imitation (*stretto*) in four parts (strings)
12. Coda of the exposition

13. Canonic imitations based on *Motif A*
    Refrain (according to the interpreter's choice).

DEVELOPMENT

14. Development of the *First Theme* followed by *Motif A*. Extremely close imitations (*Stretti*) of all instruments, based on *Motif A*
15. The *incipit* of the *First Theme* compressed in its duration (oboe)
16. *Motif A* (inverted)[2]
17. The *incipit* of the *First Theme* compressed in its duration (flute)
18. *Motif A* (inverted)
19. The *incipit* of the *First Theme* compressed in its duration (flute). *Motif A* (inverted)
20. The *incipit* of the *First Theme* compressed in its duration
21. Imitations based on *Motif A*

RECAPITULATION

22. The *First Theme*
23. The *First Theme* transposed to different tonalities
24. Imitative counterpoint based on *Motif B*
25. Imitative counterpoint based on *Motif A*
26. The *Second Theme* (in the C-tonality). This is surrounded and followed by some contrapuntal games, based on *Motif A* and *Motif B*
27. Very close canonic imitations of the *Second Theme*
28. Coda of the Recapitulation
29. Imitations based on *Motif A*

GREAT CODA

30 The *First Theme* inverted

31 The *First Theme*, the *Second Theme*, *Motif A* and *Motif B* in a great fugato with four subjects

Just to give some examples, we will present here some score-fragments that realize items 3, 10 and 31 of our scheme, respectively.

[2] A melodic line is *inverted* when its ascending intervals are transformed into descending ones, and viceversa.

3. New exposition of the *First Theme* as a fugato

10. Canonic imitations with the *incipit* of the *Second Theme*

31. The *First Theme*, the *Second Theme*, *Motif A* and *Motif B* in a great fugato with four subjects

Interestingly enough, a deeper analysis of the Symphony's finale allows us to discover a kind of "secret organization" of our four characters: the *First Theme*, the *Second Theme*, *Motif A* and *Motif B* (which will be now simply indicated by numbers 1, 2, 3 and 4, respectively). One is dealing with a *virtual* (or *latent*) polyphonic structure that represents an ideal basis for the whole composition, even if such a structure does not actually appear in the written score.[3]

[3]Also contemporary music makes sometimes use of virtual structures. As an example, we might quote a composition of one of the authors of this book. In the *Adagio* for strings in four parts, *Le tombeau perdu* (by A. R. Luciani), the first four notes of the *First Theme* of the "Jupiter" Symphony's finale (repeated on different degrees of the scale, with different

The latent polyphonic structure

values) play the role of a *secret line*, a kind of hidden counterpoint that, according to the score-prescription, should not be played.

CHAPTER 11

Tonal ambiguities

1. Beethoven's hymn of thanksgiving in the Lydian mode

The third movement (*Molto Adagio*) of Beethoven's Quartet *Op. 132* (composed in 1825) is entitled "Song of Thanksgiving to the Deity from a convalescent in the Lydian mode". The piece appears to the listener as a kind of "talisman", which seems to suggest a link between music and cure (Beethoven had just recovered after a serious illness). From a strictly musical point of view the movement can be defined as an instrumental chorale written in the Lydian mode.[1] According to our perception, accustomed to tonal music, the use of one of the ancient modes creates a vague sense of uneasiness. The ambiguity is mainly determined by the fact that the sounds of the Lydian scale belong at the same time to the F-major tonality (except the B, which is natural in the Lydian mode) and to the C-major tonality (which is, however, somehow eluded, because the gravitation upon the fundamental degrees of the scale is here missing). All this has, of course, important effects from the stylistic point of view: the listener turns out to be surrounded by a music that is archaic and modern at the same time, overcoming in this way the tonal grammar of the Western tradition. Another "miracle" that contributes to the ambiguity of the movement is due to the harmonic frame: although the structure appears spoil and essential, the counterpoint-procedure brings about a number of complex dissonances.

[1]The three following scales 1) F, G, A, B, C, D, E; 2) E, F, G, A, B, C, D; 3) D, E, F, G, A, B, C correspond to ancient ecclesiastic modes, derived from Greek harmonies and termed *Lydian*, *Frigian*, and *Doric*, respectively. The natural minor scale A, B, C, D, E, F, G, introduced by Glareanus in the 16th Century, corresponds to the *Aeolian* (or *Hypodoric*) mode, which has survived in modern tonal music. The same author has termed *Ionian* the modern major scale C, D, E, F, G, A, B.

FIGURE 1. Beethoven's hymn of thanksgiving

The presence of ancient modes is somehow hidden in many Beethoven's compositions, giving rise to tonal ambiguities. As an example, we might recall the first measures of the "Moonlight Sonata" (*Op.27 No.2*), where the Aeolian mode can be recognized in the bass-line. Also the celebrated *Bagatella Wo O 59* "For Elise" (whose structure is only superficially simple) alludes to the Aeolian mode in the arpeggio-sequence that precedes the last return to the main theme. Here the arpeggio in the *A*-minor tonality and the underlying chords that avoid the *G* sharp create a vague Aeolian colour.

2. Modulations and tonal instabilities

Tonal ambiguities have a very important role in the framework of tonal music, while tonal stability is, instead, often responsible for monotone and boring effects. As is well known, tonality-changes and tonal instabilities are generally based on a procedure that is called *modulation*. From an abstract point of view, we can represent the twenty four (major and minor) tonalities of the tempered system as some clear and well-defined *possible worlds*. For any scale, the membership-relation is exactly determined (like in classical set theory): any single note either belongs or does not belong to a given scale (*tertium non datur*!). Is it sensible to apply the classical membership-relation also to the case of musical phrases? Consider again the *First Theme* of the last movement of the "Jupiter" Symphony:

Here the first seven measures clearly belong to the main tonality in the movement (*C* major). However, a foreign element (the *C* sharp, which belongs to the *D*-major tonality) suddenly appears in the 8th measure. Is it sensible to claim that the *incipit* of the fourth movement of the "Jupiter" Symphony does not belong to the *C*-major tonality? The answer to this question is of course negative: the fleeting appearance of a *C* sharp seems to behave like a kind of *appoggiatura*, which contributes to the charm of the *First Theme*. How can we describe, from an abstract point of view, such tonal "exceptions"? One might have recourse to fuzzy-set theoretic tools, which have often been used in different frameworks in order to represent ambiguous semantic phenomena. We will choose, however, to apply (in this case also) the general notion of quantum superposition. In this perspective, the role of each tonality can be compared to a classical register, that represents a precise and certain knowledge, devoid of any form of ambiguity (see Chapter 3). As we have learnt, registers can be superposed, giving rise to quregisters: one obtains, in this way, uncertain pieces of information that assign well determined probability-values to the classical certainties represented by registers. As a consequence, each uncertain knowledge turns out to attribute different weights to the elements of a given set of certain pieces of information. In a similar way, it seems reasonable to represent a musical phrase, that does not definitely belong to a given tonality, as a superposition of different tonalities, each accompanied by a given weight. Accordingly, the *First Theme* of the fourth movement of the "Jupiter" Symphony can be described as a superposition of two tonalities: the C major (whose weight is great) and the D major (whose weight is small). Notice that we are not using the term *superposition* in the sense of *polytonality*. A typical *polytonal phrase* consists of distinct parts that belong to different tonalities and are performed at the same time. Quantum superpositions instead express some global forms of tonal ambiguity. As expected, in the case of music, sharp numerical distinctions would be quite unappropriate and useless. Apparently, it will be sufficient to refer to a "short" sequence of weights like the following: *null*, *almost null*, *very small*, *small*, *intermediate*, *great*, *very great*, *almost total*, *total*.

The characteristic role of musical modulations is creating tonality-changes. The starting point may be a certain tonality, which is followed by a situation of tonal ambiguity, where different tonalities co-exist; finally, one arrives at another tonality, which may be either closely related or distant with respect to the original tonality. One is dealing with a process that generally transforms a given *tonal order* into a different tonal order, passing through a *disordered structure*. All this seems to be quite similar to what happens in quantum computation when a *genuine* quantum logical gate is applied. For instance, a typical case is represented by the gate *square-root of negation*, whose behaviour seems to be *prima facie* quite strange. Suppose that the gate-input is a certain piece of knowledge (say, the bit 1, corresponding to the answer *YES*). Then, a first application of our gate transforms the bit 1 into a perfect *quantum perhaps*: a superposition of the two bits 1 and 0, where each bit is equally probable. Finally, a second application of the gate, transforms the *quantum perhaps* into a new certainty: the bit 0 (i.e. the answer *NO*). In this way, an *ordered informational structure* is transformed into a new ordered structure, passing through a maximum of informational disorder (which corresponds to the total uncertainty).

3. Modulations in Schubert's "Kennst du das Land"

In order to give some examples, we will consider again the *Lied* "Kennst du das Land" by Schubert, the great master of tonal instabilities, whose modulations have sometimes been called "divine". We have noticed (in

Chapter 8) that Schubert's intonation of Goethe's poem seems characterized by a "consoling" element and by a positive impetus, which co-exist with a shadow of doubts and of anxiety, generally expressed by the piano-accompaniment.

We will now go more deeply into our analysis, focusing on the tonal relationships in the first stanza of the *Lied.*[2] As often happens in tonal compositions, the *incipit* belongs to a certain tonality, without any shadow of ambiguity: in this particular case, one is dealing with the A-major tonality. The melodic line of the singing is developed in the framework of this clear and sharp tonal world during the first five measures (*Kennst du das Land wo die Zitronen blüh'n, im dunklen Laub die*). At the same time, already in the second measure, the piano is suggesting a little doubt, by means of a foreign note (an E sharp, which belongs to the corresponding minor tonality). This creates a major-minor superposition, where the minor tonality has a very small weight. The 6-th measure (*Gold-o-*) realizes a sudden modulation (from the A major to the C major) which concerns both the voice and the piano. Soon after the C-major tonal order is rigorously preserved from measure 7 to measure 10 (*-rangen glüh'n, ein sanfter Wind vom blauen Himmel*). Measure 11 gives rise to a new tonal ambiguity, modulating to the F-major tonality. From the expressive point of view, this corresponds to the most serene and bright fragment of the first stanza (*die Myrte still, und hoch der Loorber steht?*). But then the piano (solo) confuses the situation again: measures 15 and 16 are characterized by a fluctuating tonal uncertainty, ending in an arpeggio on the dominant-seventh chord, which is prolonged by a *hold.* The tonal oscillation seems to express Mignon's psychological confusion, whose recollections about the place where she wishes to return are fairly obscure. Her question *Kennst du es wohl?* remains suspended above an empty space, on the fifth degree of the E-major tonality, after the piano-chord (in measure 18). The second part of the first stanza (*Etwas geschwinder*) is characterized by an increasing tonal uncertainty. In measures 19-23, the piano realizes a sequence of chromatic scales, performed both by the right and by the left hand. Apparently, using a chromatic scale (where the interval between two contiguous elements is constantly one semitone) determines a maximum of tonal uncertainty. One is dealing with a situation that seems to represent a quantum-like superposition of all possible tonalities. We might compare such a musical pattern to the case of an uncertain quantum information (quregister) that assigns one and the same probability-value to all possible certain pieces of information (registers). Measure 25 realizes a fleeting return to the main tonality (A major), upon the words *mein Geliebter*. Soon after, however, the piano-chromaticism starts again (in measures 26-28). This seems to create a contraposition between the bright certainty of Mignon's feeling and her obscure recollections. There is a moment where her strong will and desire

[2]See the *Lied*'s score in the Appendix.

seem to supervene: measure 29 modulates to a new tonality, the *D* major, which is rigorously preserved during 6 measures (*dahin, dahin möcht' ich mit dir, o mein Geliebter, zieh'n*). Here the joyful triplets performed by the piano (in a dynamic *crescendo*) seem to evoke the sound of an approaching postillion (a kind of musical realization of the Doppler-effect). The stanza's finale is based on a "divergence" between the singing and the piano. The voice pursues its positive impetus, in the frame of a tonal structure that is common to the *D* major and to the *A* major tonalities. The piano instead proposes again a chromatic development that turns out to be more and more ambiguous and dissonant. Only the last measure realizes a resolution, with a perfect tonic chord in the main tonality.

In many Schubert's *Lieder* the piano has the role of expressing some obscure (possibly unconscious) thoughts by means of ambiguous tonal structures, while the voice follows a clear tonal line. Let us think, for instance, of the celebrated *Lied* devoted to another great character created by Goethe: *Gretchen am Spinnrade* (Gretchen at the spinning wheel), from *Faust* (Fig. 2). Schubert composed this *Lied* in 1814, one year before Mignon's *Lied*. In this case the piano has also a descriptive function: the sextuplets that are obsessionally repeated seem to evoke the sound of the running spinning wheel. At the same time, the sudden appearance of dissonant chords fully expresses Gretchen's emotion. For instance, when the voice, perfectly in tune, sings "und ach, sein Kuss!" ("and ah! his kiss!"), the piano performs a *diabolus in musica* (a chord consisting of two contiguous minor-third intervals). And the "diabolic" dissonance occurs soon after again, always accompanying the idea of a kiss (*und küssen ihn*). Only sorrow and loss of peace are characterized by a perfect tonal certainty: *Meine Ruh ist hin, mein Herz ist schwer!* (My peace is gone, my heart is heavy!).

4. Indeterminacy and infinity in *Tristan und Isolde*

In many compositions that belong either to the classical or to the Romantic tradition tonal ambiguities often correspond to transition-moments: the musical structure is generally based on alternating forms of tonal certainty and of tonal uncertainty. This is the basic reason why it makes sense to compare such musical patterns with quantum logical gates, which possibly transform certain pieces of information into other pieces of information that are totally uncertain (and viceversa).

A complete tonal indeterminacy is instead the dominating structure of Wagner's *Tristan und Isolde*, where any form of tonal certainty seems to vanish, while tonal superpositions are preserved during a very long time, without any resolution. The *Prelude* (which Wagner wrote with fluent facility before composing the three Acts of the opera) sums up and anticipates all the basic elements of *Tristan und Isolde*: the musical ideas (represented by the different *Leitmotive*), the melodic, harmonic and rhythmic aspects,

FIGURE 2. Gretchen at the spinning wheel

the orchestration. Tonal instability is apparent already in the first measures performed by the cellos:

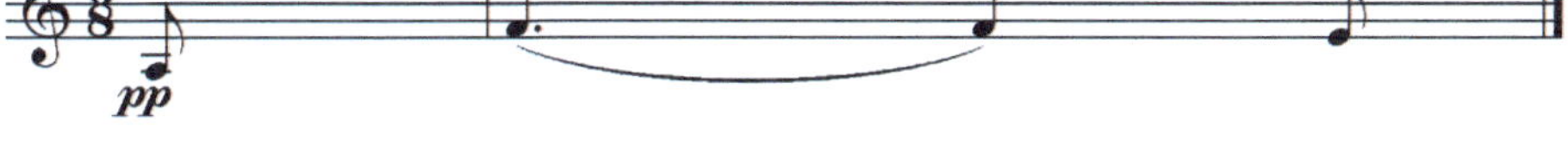

One is dealing with a minor-sixth melodic interval (A- F), that does not belong to the triad of the main tonality of the *Prelude* (A minor). This immediately creates an exasperated and prolonged harmonic tension, which induces us to desire the fifth degree of the scale (the E). The second measure contains the first appearance of the famous "Tristan-Akkord", based on the superposition of two different fourth-intervals: the "diabolic" *augmented fourth* (F-B) and a *perfect fourth* (D sharp - G sharp):

A melodic line starts soon after, based on *appoggiaturas* (minor-second intervals), which are repeated several times on higher and higher notes that are more and more distant from the initial tonality. The chromaticism, connected with the *desire-theme* creates the illusion of an infinite melody, without interruption after the initial uncertainty. At the same time, the lack of perfect cadences seems to allude to a sequence tending to a limit that is in fact inaccessible.

Music expresses in this way a psychological itinerary from the first desire-explosion to an overwhelming love's confession, through a continuous and vane *crescendo*, where any satisfaction causes a new desire, until the "death's joy" in the final redemption (Fig.3). Only in the finale of the opera - after Isolde's last words "ertrinken, versinken, unbewusst, höchste Lust!" - we have a musical resolution of all forms of chromaticism and dissonance through a perfect tonic chord (in the B-major tonality).

Tristan und Isolde's ambiguities have been significantly commented by one of the greatest interpreters of Wagner, the conductor Daniel Barenboim, who has realized some memorable performances of the opera[3]

> *Ambiguity is, in my opinion, the key-element of this opera. An ambiguity that concerns the characters, the situations, the different interpretations of past events, the premonitions about the future. Wagner's genius has invented an ambiguous language that is characterized by chromaticism. In this score there is no direct and definitive harmonic resolution: we are always waiting for a resolution that we never reach. The lack of a full resolution in harmony means preserving some dissonant elements, which allows us to imagine different possible resolutions, while no definite resolution is actually proposed..... Such ambiguity of the musical language,*

[3]For instance, in Bayreuth in 1995 and in Milan in 2007.

FIGURE 3. Tristan and Isolde

combined with the characters' ambiguity, is transformed into a strong tension that appears unbearable..... Such ambiguity, (which is, in fact, also mutability) contributes to change the feelings of the characters, as well as our feelings as performers or listeners. The study of Tristan und Isolde *has changed our vision of things, our frame of reference, our valuation that does not concern music only, since it involves perhaps life itself.*[4]

[4]D. Barenboim, P.Chéreau, *Dialoghi su musica e teatro. Tristano e Isotta*, Feltrinelli, Milano, 2008, pp. 15–16. Translation of the authors of this book.

CHAPTER 12

Psychological and narrative ambiguities

As we have seen in the previous Chapters, the concept of superposition represents an useful abstract tool that allows us to investigate different forms of musical ambiguity. The dramatic structure of lyric operas is often based on ambiguous situations that concern both the narrative form and the character-psychology. In this sense, musical theatre seems to have many properties in common with prose-theatre. There is however an important difference: the interaction text-music, which is a characteristic of lyric operas, naturally creates a kind of many-dimensional abstract space, where different antagonistic elements may be simultaneously present. The fragment of the duet between Violetta and Germont in *La Traviata* (mentioned in Chapter 6) is a typical case where music expresses a co-existence between contradictory thoughts and feelings, according to a parallelism that represents one of the specific aspects of our mental processes. This might perhaps explain (at least to a certain extent) one of the reasons why music is a kind of universal thought, which seems to be deeply connected with the sphere of our emotions. A paradigmatic example is represented by Wagner's operas, where the *Leitmotiv*-technique creates some very special effects, bringing into light the similarities between the "flowing" of musical designs and the developing of our (conscious or unconscious) thoughts, through associations, recalls and continuous variations. One is dealing with some deep cognitive phenomena that are today investigated in the framework of neuro-sciences.

Psychological ambiguity has a very important role in many lyric operas, where the light and shade about the dichotomies "good-evil", "love-hate", "tragic-comic" occur with infinite variations. As happens in a number of literary works, a great character is often described as a superposition of antagonistic properties, which differently emerge in different situations; in contrast with most "minor" literature, where human persons are usually represented as sharply one-dimensional: either good or wicked, either heroic or coward, without any fuzzy shade (we need only think of serial novels, cartoons and so on).

One of the greatest Verdi's characters, who wonderfully realizes the contraposition "good-evil" is Rigoletto. Interestingly enough, Verdi describes Rigoletto as perfectly aware of his psychological duality. A significant example can be found at the end of the first Act of the opera, when Rigoletto returns to his daughter Gilda, after a violent struggle with Monterone, who had cursed him for his cruelty. He sings:

Odio a voi cortigiani schernitori!
Quanta in mordervi ho gioia!
Se iniquo son, per cagion vostra è solo.
Ma in altr' uomo qui mi cangio![1]

And his self-consciousness moment (*Ma in altr' uomo qui mi cangio!*) is musically preceded by a sweet melody performed by the flute, which anticipates, *pianissimo*, the line of his singing. Significantly enough, the phrase terminates with a kind of *deceptive cadence* [2]:

1. Ambiguity and contradictions in *The Magic Flute*

An opera that deeply exalts both psychological and narrative forms of ambiguity is *The Magic Flute* by Mozart. At a first listening (or at a first reading of the libretto) one might have the impression that the dramatic structure is similar to a detective story, where one intentionally tries to deceive the reader, by continuously interchanging the roles of "innocent" and "guilty", "good" and "wicked" persons. The most intriguing character of this ambiguous game is *The Queen of the Night*. Her first aria creates for Tamino the false image of a kind of *mater dolorosa*:

[1] My hate upon you, sneering courtiers/How I enjoy snapping at your heels!/ If I am wicked the fault is yours alone./ But here I become another person.

[2] One uses to speak of *deceptive cadence* when the harmony, instead of concluding upon the tonic, against what is expected, leads to some other degree of the scale.

FIGURE 1. The ambiguity of The Queen of the Night

Zum Leiden bin ich auserkoren,
denn meine Tochter fehlet mir;
durch sie ging all mein Glück verloren:[3]

The musical form is a *Larghetto* in *G* minor, a tonality that is often used by Mozart to express sadness and melancholy. This first image of the Queen is however immediately unveiled (but only musically!) by the virtuoso-acrobatic ornaments of the voice, dealt with as an instrument that turns out to freeze any emotion (Fig. 1). At the same time, the text insists in transmitting a positive message:

und werde ich dich als Sieger sehen,
so sei sie dann auf ewig dein. [4]

.

The Three Ladies also seem to play a quite ambiguous game, which is apparent in the first terzetto at the very beginning of the opera: their falling

[3]I am chosen for suffering/ For my daughter is gone from me;/ Through her all my happiness has been lost.

[4]And if I see you return in triumph,/ Then she will be yours forever.

in love with Tamino may deceive the listener, and also their singing about "Liebe" (love) and "Bruderbund" (brotherhood) appears to be absolutely sincere, without any fiction-shadow. Furthermore, strangely enough, the magic flute, which will represent for Tamino a guide and a liberation-tool is described as a gift offered by the *Reign of the Night*. One can wonder whether the intended suggestion is that sometimes negative situations may give rise to positive effects. Act II only will clarify that the Ladies and their Queen share the same wicked intention: a perfect "good-evil" superposition, where sometimes text and music seem to follow different semantic paths.

Tamino himself is not devoid of some ambiguous features. In principle, he should represent the character of a brave young man in love, who tries to rescue his beloved girl from the evil's menaces. However, at the very beginning of the opera he is presented in a quite non-heroic way. His first words are a help-request:

Zu Hilfe! zu Hilfe!
sonst bin ich verloren! [5]

And those who finally save him from a horrible serpent are three young women: the Ladies of the Queen themselves. Is there any other opera where the positive hero, the tenor in love is described in a similar way? At first sight we might suspect that one is dealing with an ironic making fun of serious operas. However, this is not the case. Any parody-impression is immediately contradicted by Tamino's first aria: an intense and moving song that seems to anticipate the style of Romantic *Lieder*.

Dies Bildnis ist bezaubernd schön,
wie noch kein Auge je gesehn!
Ich fühl es, ich fühl es,
wie dies Götterbild
mein Herz mit neuer Regung füllt. [6]

.

Even the pair Papageno-Papagena, who represent the most naive characters of the opera, gives rise to some ambiguous situations. Papageno, first appearing in the *Reign of the Night*, has an ambiguous identity, which is expressed both by his singing and by his dress: he seems to be at the same time a bird and a human person. Papagena is ambiguous too: she is first presented as an ugly old woman, who suddenly disappears, scaring the poor Papageno. But finally she will turn out to be a deeply desired aim, a kind of *alter ego* for Papageno, who will try to commit suicide for her.

The listener is introduced into the enlightened reign, Sarastro's temple, in the most deceptive way: Pamina has been kidnapped and the lustful

[5] Help! Help!/ Otherwise I am lost!

[6] This image is enchantingly lovely,/ Like no eye has ever beheld!/ I feel it as this divine picture,/ Fills my heart with new emotion.

Monostatos has attempted sexual violence against her. One has often discussed about the role of Monostatos in the opera. What is doing a person of that kind in a world that is dominated by the reason's ideals, by the research of a universal peace and by the refusal of any vengeance? Perhaps the message that is implicitly transmitted is the following: any progressive social project, any utopia is unavoidably accompanied by some negative elements.

There are some contingent events in the history of *The Magic Flute*'s composition that may perhaps explain some contradictions of the libretto. The original project that Emanuel Schikaneder (writer, actor, singer and first interpreter of the role of Papageno) had proposed to Mozart belonged to the *Zauberposse*-genre (magic comic opera): there were a good fairy, a cruel magician who kidnapped the fairy's daughter and some popular comic characters like a *Vogelmensch* (a bird-man), a funny person who did a lot of disasters, but finally succeeded in reaching his goal. Mozart began to compose the music for this story with great enthusiasm. But suddenly perturbing news arrived: the *Marinellis Theatre* had just represented the première of a *Zauberposse* whose libretto was quite similar. Schikaneder proposed to interrupt their own work. But Mozart, who had already composed a lot of music, did not agree. Hence the decision was to change the plot: the good fairy becomes cruel, while the magician is transformed into a wise man. This is the origin of the character of Sarastro, with the idea of using themes and rites of the Masonic world.

Paradoxically, all these contingent events have positively contributed to the final opera's result. Without the competitors' interference, *The Magic Flute* would probably have been a kind of naive fairy tale with a wonderful music. What instead has been created is a deeply enigmatic construction, where ambiguity represents an essential feature and the main part of the obstacles to be solved in the initiation-process. Its fascinating character also depends on the different interpretation-levels that are suggested by the opera: one is dealing at the same time with a fairy tale, a comedy, an *exemplum*, where tragic and naive features wonderfully co-exist. An important role is played by different forms of symbolism, which have been suggested both by the Masonic world and the by the exoteric thought about music: from Pythagoras to Plato and to neo-platonic philosophers, from science to alchemy.

2. Ambiguity and indefiniteness in Pelléas et Mélisande

Although ambiguity is a characteristic cipher of *The Magic Flute*, in the opera's finale every form of uncertainty is solved: the "good" is separated from the "evil" and all characters take on clear and defined roles. If we want to continue in using a quantum language we might say that at the end there is a sort of *collapse of the wave-function*: the superpositions disappear and are transformed into precise and sharp pieces of information.

There is instead a musical work from the 20th Century that seems based on an absolute form of ambiguity, where superpositions are never solved: it is Debussy's *Pelléas et Mélisande* (first performed in 1902 at the *Opéra Comique* in Paris). The concept of this opera seems totally different from any operatic tradition heretofore, both from a poetic point of view and from a musical one; indeed it provoked burning controversies even before its first performance. Maurice Maeterlinck's play, which provided the libretto, had already excited opposed reactions among the intellectuals and critics present at the first Parisian performance in 1893. The Lugné Poe theatrical troupe put on the play using costumes inspired by Pre-Raphaelite art and with a shadowy lighting in grey tones, separating the stage from the audience with a light tulle curtain: every expedient was used to accentuate the sense of mysterious blurring and create temporal and spatial distance from the subject, whose aesthetic substance deeply impressed Debussy.

Maeterlinck's Symbolist poetic dwelt on two fundamental ideas:

- a particular sense of death, which ambiguously represents at the same time the only real certainty in anyone's life and an absolute mystery up to the moment that death itself is experienced.
- Destiny as a "sublime character", which ineluctably governs the course of life and to which humans can only submit themselves.

Curiously, ambiguity was a strong element even in the personal relationship between Maeterlinck and Debussy. Immediately after attending the play's performance, the composer started to work incessantly on the musical production and in about three years time he would pronounce finished his version for voice and piano. At the same time, through friends Henri de Régnier and Camille Mauclair, Debussy did his best to secure Maeterlinck's permission to set the play to music. This was his second attempt to collaborate with the Belgian writer, after having in vain requested the exclusive rights to set an opera to the play *La Princesse Malaine*, another work with Gothic influences that had fascinated the composer for its Symbolist aesthetics, but that Maeterlinck preferred to entrust to Vincent d'Indy. Worried about the possibility of a second refusal, in 1893 Debussy took his friend Pierre Louÿs to Ghent to his meeting with the poet, who welcomed Debussy enthusiastically, impressing him by the width of his culture and his generous availability, leaving the musician free to make all the cuts that he thought necessary and even suggesting others. Lockspeiser tells us that Louÿs was nonetheless deeply embarrassed because neither the poet nor the composer dared speak openly, so that it was up to him to talk now to one, now to the other. After the visit, relations between Maeterlinck and Debussy went no further until a ferocious argument suddenly broke out prior to the opera's première in 1902. Only after *Pelléas et Mélisande* was officially accepted by the *Opéra Comique* did Debussy start arranging the orchestration. When the opera was complete, pressure from the composer's friends to reveal the work were such that Albert Carré, director of the *Opéra Comique*, decided to risk putting it on in repertoire, on the advice of the conductor André Messager. When Carré and Debussy were agreeing on the choice of singers, they both decided to give the role of Mélisande to the Scottish soprano Mary Garden. However, at this point Maeterlinck re-emerged from the silence and disinterest he had shown with regard to the staging of the opera and demanded that the role of the female protagonist be played by the actress (and his lover) Georgette Leblanc, on the basis of an agreement that he claimed he had made beforehand with Debussy. Maeterlinck's violent quarrel with the composer even resulted in threats and the challenge to a duel, as well as in a letter written to *Le Figaro* a fortnight before the première, in which the author hoped that the opera would turn out to be a memorable fiasco. However, all this failed to compromise its enormous success. Maeterlinck refused to attend a performance until 1920, when, two years after Debussy's death, he was obliged to admit the great value of the opera.

Some experts have suggested that the true motive for Maeterlinck's attack was the fact that in *Pelléas et Mélisande* the music overwhelmingly underlines and strengthens the symbolism of the text. By virtue of its ambiguous vagueness the play constituted the ideal libretto for Debussy, with all the issues that he had already expressed to his composition teacher Ernest Guiraud, in 1889: that is, a text which, "by half saying things", left the musician with the possibility to amplify through sound the emotions kindled by words ("it will allow me to graft my dream onto his"); a play whose characters and whose story "do not belong to a time or a place", whose scenes are "mobile, different in place and nature; in which the characters do not fight, but endure life and fate". With these words, that prefigure what will be his *Pelléas*, Debussy is expressing an ideal of dramatic music that agrees with Mallarmé's aesthetics precisely in its necessary margins of vagueness on which the evocative capacity of Symbolist art feeds. For Debussy music begins where words are not able to give expression:

> *I am not tempted to imitate what I admire in Wagner. I conceive a different dramatic form: music in itself begins where words are not able to give expression; it is made for the inexpressible. I would like music to have the air of coming out of the shadows and, gradually, returning; and being always discreet.*

Pelléas et Mélisande's plot is based on nothing, there are no precise or concrete facts and all events occur in an atmosphere of uncertainty. Mélisande always evades the questions put to her and says nothing of herself, apart from her name. Golaud, her husband, suspects her of being unfaithful, but he will never find proof, despite trying to obtain it through his small child Yniold, who in his infantile ingenuity can never be a reliable informer. The child-like fondness exchanged by Pelléas and Mélisande might seem to the audience to be an innocent joke, but Mélisande dies giving birth to a daughter whose paternity is uncertain. Everything is based on misunderstanding and vagueness, fate drags the characters along without an explanation: both Mélisande and Pelléas have ingenuous and pure souls, on which the ineluctability of death is inflicted with an almost sadistic cruelty. Every character in the opera seems to live in a vague set of "possible parallel histories". From this point of view it is not unnatural to draw a similarity with a particular approach to quantum mechanics that has been called *The consistent-histories theory* where the temporal evolution of physical objects is represented as a sheaf of possible *histories*, with respect to which both past and future are equally indeterminate.[7]

The relationship between words and music in *Pelléas* is itself marked by ambiguity in a subtle choice: Debussy left Maeterlinck's text in prose; his sensitivity to the sound of words oriented his choice of a continuous syllabic recitation, forgoing direct melodizing ("I do not want to risk my music being whistled down the boulevards of Paris"). The hazard of monotony is completely avoided by the extremely fanciful orchestration that supports the vocal score, which in turn frequently expands into passages of vibrant lyricism. Although it does not conform to any particular type of vocal style, the strong stylistic unity in *Pelléas et Mélisande* derives from the lack of interruptions in the phrases and the capacity to maintain, at the same time, a continuous and flexible musical tension, that constantly tries to merge the lyrical flow with the exigencies of dramatic art. The scenes of the opera's five Acts are like *tableaux* that illustrate "photograms" of the action and static situations, separated out by an unspecified timing that runs along with the orchestral interludes. Musical and psychological unity is also ensured by the use of *Leitmotive* which, although owing much to Wagner's musical drama, bear no reference to the declamatory symphonic nature with which Wagner re-proposed orchestral themes alongside the relative narrative and symbolic information.[8] The recurrent themes of *Pelléas et Mélisande* are used with subtle discretion, thanks to the orchestral transparency and the extraordinary imagination with which they run through each other by intersections, associations and underlying similarities. These motifs, which ensure great depth and significance to the dramatic action, are often fragmentary, changing according to the situation or mood of the characters that they are associated with. Sometimes they suggest reality, sometimes only an idea of reality, because, more than the sensitive side of phenomena, the musician is interested in their association to thoughts and sentiments, seen through the fleeting vagueness of their melodic and expressive profile. Through the *Leitmotive* in *Pelléas et Mélisande* Debussy recreates a new symbolism of the musical language, distancing himself from Wagner and the models of musical Romanticism above all, where the "eternal" ("Tristanian") themes of love

[7]See, for instance, M. Gell-Mann, *The Quark and the Jaguar: Adventures in the Simple and the Complex*, W. H. Freeman and Company, New York, 1994.

[8]See M. Emmanuel, *Pelléas et Mélisande de Claude Debussy: étude historique et critique*, Librairie Delaplane, P. Mellottée, Paris, 1926.

FIGURE 2. Mélisande

and death are at stake: themes that usually provide the pretext to reinforce expression by all means and push the orchestra towards delirium, underlining transports of passion, pain and desperation. On the contrary, the striking first scene of Act III - the dialogue between Pelléas and Mélisande at the castle's tower window, with her hair falling down around Pelléas - is sustained for its duration by the orchestra playing *pianissimo*, despite the deep emotional tension (Fig. 2).

In the second scene of Act IV - when, near the fountain, the two protagonists declare their love - the orchestra becomes suddenly silent, precisely at the words "Je t'aims" and "Je t'aims aussie", while until then it had been playing loud and *crescendo*. It will re-emerge later, from the shadows, with a *pianissimo* (*ppp*) to accompany Pelléas' passionate recitative.

In this opera, a masterpiece of paradoxes, we find silences that produce sound and sounds hidden within silences. Stefan Jarocinski maintains that "almost all Debussy's music emerges from silence, gradually disappears and falls back into silence; the composer seems to be listening to the mysteries of life, death and the 'supersensible' ". According to Debussy himself silence is "perhaps the only way of putting the emotional value of a phrase into relief".[9] The eloquent silences scattered throughout *Pelléas* should be considered a musical structure equivalent to the sound of an opera that is more full of absences and evocations than of "full-bodied" characters: everything within it is fairy-tale-like, allusive and inconsistent and it is dominated, without opposition, by the far distance of horizons and time. Thus, the kingdom of Allemonde, where the story is set, contains, even within its name, the semantic allusion to a place that could be anytime and anywhere. There are six characters effectively present on stage (Golaud, Mélisande, Pelléas, Arkel, Geneviève and Yniold), but many others are evoked; even the generational relationships between them are enveloped in an intentional haziness; the wife of Arkel and mother of Geneviève, the latter's two husbands (with one of whom she had Golaud, the other Pelléas), Golaud's first wife (mother of the little Yniold) and the father of the child Mélisande gives birth to remain a mystery. The old king Arkel, with his dark and threatening timbre, conditions the scenes in which he appears with the sense of the imminence of death and the unavoidability of destiny to be undergone, tainting the castle and the whole landscape of Allemonde with the shadow of weariness and old age. The forest surrounding the castle oppresses Mélisande's spirits with its tomb-like obscurity - as it had oppressed Geneviève's in her time - and causes all the characters to long for light and clarity. Mélisande, who we imagine to be hardly more than a child due to her infantile language and her ingenuity, comes into the female typology of Pre-Raphaelite iconography. This is above all because of the lure of her long hair, which enchants first Golaud and then Pelléas, and grows into a cascade falling from the tower window or on the fountain's mirror of water; she is a child-woman who finds in Pelléas something of a playmate. While he has nothing in common with the powerful virility of Golaud, a man who reasons like a soldier and hunter, incapable of understanding Mélisande's melancholy and delicate psychological transparency. As far as feelings are concerned, Geneviève is the only clear and defined character of the whole story: she is a mother and, as such, has the function of bringing light and being a guide. The omnipresence of the element of water, in the form of the fountain, the sea and Mélisande's flowing hair as well as the frequent verbal and musical references to water bring to mind its identification (not only "Freudian") with the maternal

---

[9] See S. Jarocinsky, *Debussy. Impressionnisme et symbolisme*, Éditions du Seuil, Paris, 1970.

figure: like the constant tension towards the light, it constitutes a hidden thread throughout the opera.

The recurrent themes of light, shade and the liquid element correspond in the orchestra to a complex tissue of allusions, thematic refrains and *chiaroscuro* - tone colours that unfold in the delicate weave of an instrumentation of camera-like transparency. Debussy subdivides the instruments, isolating them even within single orchestral sections, and creates the impression that each of them is also treated like a soloist. The harmonic progressions of the *Pelléas et Mélisande* score are also an example of ambiguity, with their frequent enharmonic passages and the chords used not only in the functional sense, but also in the sense of colour and timbre. With the frequent weaving of different rhythmic figures, these elements contribute to the creation of a strikingly original composition-technique, based on a kind of indetermination-principle, that gives the impression of having been developed by instinct and that will inaugurate a new chapter of musical theatre in the 20th Century. As the conductor of the opera's first performance André Messager said, when in the final scene, the dying Mélisande asks that the window be opened, through it comes not only the light of sunset, but modern music too.

Appendix

Mignon's song by Schubert

Etwas geschwinder.
Da - hin, da - hin! da - hin möcht' ich mit dir, o mein Ge-lieb-ter, ziehn; da - hin,
Da - hin, da - hin! da - hin möcht' ich mit dir, o mein Beschüt-zer, ziehn; da - hin,
da - hin, da - hin! da - hin möcht' ich mit dir, o mein Ge - lieb - ter, ziehn, da - hin, da - hin, da - hin, da - hin!
da - hin, da - hin! da - hin möcht' ich mit dir, o mein Be - schüt - zer, ziehn, da - hin, da - hin, da - hin, da - hin!
cresc.
fp
ff

Wie oben.
Kennst du den Berg und sei-nen Wol-ken-steg? Das Maulthier sucht im Ne-bel
sei-nen Weg; in Höh-len wohnt der Dra-chen al-te
Brut; es stürzt der Fels und ü-ber ihn die Fluth,
kennst du ihn wohl?
Etwas geschwinder.
Da-hin, da-

hin! da - hin geht un - ser Weg! o Va - ter, lass' uns
zieh'n! Da - hin, da - hin, da -
hin! da - hin geht un - ser Weg! o Va - ter, lass uns
zieh'n! Da - hin, da - hin, da -
hin, da - hin!

Mignon's song by Schumann

hin möcht ich mit dir, o mein Ge-lieb-ter, ziehn, da-
dimin.
p
hin, da- hin mit dir, o mein Ge-lieb-ter, ziehn.
dimin.
p
p
Kennst du das Haus? Auf Säu-len ruht sein Dach, es
Ped.
Ped.
cresc.
glänzt der Saal, es schim-mert das Gemach, und Marmor-bil-der stehn und
cresc.
Ped.
Ped.
www.everynote.com
sehn mich an: was hat man dir, du ar- - - mes
fp
fp

cresc.
Kind, ge - tan? Kennst du es wohl, kennst du es
fp
cresc.
Ped.
f
wohl? Da - hin! da - hin möcht ich mit
f
Ped.
dimin.
p
dir, o mein Be-schü-tzer, ziehn, da - - hin, da -
dimin.
p
3
hin mit dir, o mein Be-schü-tzer, ziehn.
www.everynote.com
p
Kennst du den Berg und sei-nen Wol-kensteg? Das Maul - tier sucht im Ne-bel sei-nen Weg;
Ped.
Ped.
Ped.
Ped.

cresc.
in Höh - len wohnt — der Dra - chen al - te Brut, es
cresc.
fp
cresc.
stürzt der Fels und ü - - - ber ihn die Flut. Kennst du ihn
fp
fp
cresc.
wohl, kennst du ihn wohl? Da - - hin, da -
f
hin geht un - ser Weg! o Va - ter, laß uns ziehn, da -
dimin.
p
hin, da - hin geht unser Weg, o Va - ter, laß uns ziehn!
www.everynote.com
dimin.
p
pp

Bibliography

[1] D. Barenboim, P.Chéreau, *Dialoghi su musica e teatro. Tristano e Isotta*, Feltrinelli, Milano, 2008.

[2] G. Birkhoff and J. von Neumann, "The logic of quantum mechanics", *Annals of Mathematics* **37**, pp. 823–843, 1936.

[3] M. L. Dalla Chiara, R. Giuntini, R. Greechie, *Reasoning in Quantum Theory*, Kluwer, Dordrecht, 2004.

[4] M. L. Dalla Chiara, R. Giuntini, R. Leporini, "Logics from quantum computation", *International Journal of Quantum Information* **3**, pp. 293–337, 2005.

[5] M. L. Dalla Chiara, R. Giuntini, R. Leporini, "Compositional and holistic quantum computational semantics", *Natural computing* **6**, pp. 113–132, 2007.

[6] M. L. Dalla Chiara, R. Giuntini, E. Negri, "From quantum mechanics to music", *Advanced Science Letters* **1**, pp. 169–178, 2008.

[7] M. L. Dalla Chiara, R. Giuntini, R. Leporini, G. Toraldo di Francia, "Quantum computational logics and possible applications", *International Journal of Theoretical Physics* **44**, pp. 44–60, 2008.

[8] M. L. Dalla Chiara, R. Giuntini, A. Ledda, R. Leporini, G. Sergioli, "Entanglement as a semantic resource" *Foundations of Physics* **40**, pp. 1494–1518, 2009.

[9] A.R. Damasio, *Feeling of What Happens: Body and Emotions in Making of Consciousness*, Harcourt, 1999.

[10] I. Bent, W. Drabkin, *Analysis*, Macmillan Publisher LTD, London, 1980.

[11] A. Einstein, B. Podolsky and N. Rosen, "Can quantum-mechanical description of reality be considered complete?", *Physical Review* **47**, pp. 777–780, 1935.

[12] M. Emmanuel, *Pelléas et Mélisande de Claude Debussy: étude historique et critique*, Librairie Delaplane, P. Mellottée, Paris, 1926.

[13] R. P. Feynman, "Simulating physics with computers", *International Journal of Theoretical Physics* **21**, pp. 467–488, 1982.

[14] R. P. Feynman, "Quantum mechanical computers", *Optics News* **11**, pp. 11–20, 1985.

[15] R. P. Feynman, R. B. Leighton, M. Sands, *The Feynman Lectures on Physics*, Addison Wesley, London, 1969.

[16] R. Gilmore, *Alice in Quantum Land. An Allegory of Quantum Physics*, Sigma Press, Wilmslow, 1994.

[17] M. Gell-Mann, *The Quark and the Jaguar: Adventures in the Simple and the Complex*, W. H. Freeman and Company, New York, 1994.

[18] S. Jarocinsky, *Debussy. Impressionnisme et symbolisme*, Éditions du Seuil, Paris, 1970.

[19] L. M. Parsons, J. Sergent, D. A. Hodges, P.T. Fox,"The brain basis of piano performance", *Neuropsychologia* **43 (2)**, pp. 199–215, 2005.

[20] M. A. Nielsen, I. L. Chuang, *Quantum Computation and Quantum Information*, Cambridge University Press, Cambridge, 2000.

[21] R. Penrose, *The Emperor's New Mind*, Oxford University Press, Oxford, 1989.

[22] R. Penrose, *Shadows of the Mind*, Oxford University Press, Oxford, 1994.

[23] N. Rosen,"Quantum mechanics and reality", in P. Lahti and P. Mittelstaedt (eds), *Symposium on the foundations of modern physics*, World Scientific, Singapore, 1985.

[24] A. Schönberg,"Das Verhältniss zum Text", *Der Blaue Reiter*, München, 1912. English translation of the author in *Style and Idea*, Philosophical Library, New York, 1950.

[25] G. Toraldo di Francia, "Deep Blue and the strategy of human thought", *Foundations of Science* **4**, pp. 83–93, 1999.

[26] G. Toraldo di Francia, "Music and Science", in M. L. Dalla Chiara et al. (eds.) *Language, Quantum, Music*, Kluwer, Dordrect, 1999.

[27] J. von Neumann, *Mathematische Grundlagen der Quantenmechanik*, Springer, Berlin, 1932. English translation: *Mathematical Foundations of Quantum Theory*, Princeton University Press, Princeton, 1996.

ND - #0052 - 110926 - C32 - 234/156/7 - PB - 9781848900615 - Gloss Lamination